Plant Stress from
Air Pollution

Plant Stress from Air Pollution

Michael Treshow PhD

Professor of Biology
University of Utah, USA

and

Franklin K. Anderson PhD

Senior Environmental Scientist
Utah, USA

John Wiley & Sons

Chichester · New York · Brisbane · Toronto · Singapore

Copyright © 1989 by John Wiley & Sons Ltd.

Reprinted with corrections January 1991
Reprinted June 1991

Library of Congress Cataloging-in-Publication Data:

Treshow, Michael.
    Plant stress from air pollution / Michael Treshow and Franklin K.
Anderson.
    p.   cm.
    Bibliography: p.
    Includes index.
    ISBN 0 471 92374 5
    1. Plants, Effect of air pollution on.   2. Plants, Effect of
stress on.   I. Anderson, Franklin K.   II. Title.
QK751.T74   1989
581.5'222—dc20   89–9070
                                                                   CIP

British Library Cataloguing in Publication Data:

Treshow, Michael
    Plant stress from air pollution.
    1. Plants.   Effect of pollutants of atmosphere
    I. Title.   II. Anderson, Franklin K.
    581.5'222

    ISBN 0 471 92374 5

Typeset by Photo·Graphics, Honiton, Devon
Printed and bound in Great Britain by
Biddles Ltd, Guildford and King's Lynn

Contents

Preface

We are going to assume that all readers of this book do not have a wealth of background in air pollution biology, or even plant stress. We are also going to assume that more still have not had the opportunity to be present for the discovery of the many fascinating historic milestones in this field. So we shall begin at the beginning and reveal each landmark discovery as it became unravelled and shall look at how this provided a foundation for subsequent research leading to our present understanding of the effects of air pollution on plants.

An enormous body of knowledge about air pollutant effects on plants has been garnered in recent decades. It can be found throughout the scientific literature, in symposia, the public press and in specialized books. Many excellent works are available to the specialist. Fewer are available to provide the broader background in which many of us are interested.

In an effort to provide an objective overview of air pollution biology, we shall review the major research that led to our current state of knowledge, and fully discuss the current state of our science. Air pollution biology is too vast a field to highlight every facet. Thus, attention will be directed to the more salient aspects, highlighting the more significant as well as less frequently addressed topics. The literature reviewed for this book is comprehensive and representative, but it is not exhaustive. Many excellent studies have had to pass without mention.

The opening chapter provides some botanical background and addresses the origins of air pollution. Problems began in the cities, most notably when they were based on industry. Lichens were the first organisms noticeably affected by pollutants, and the volume next concerns them and the research directions to which they led. Sulfur dioxide from coal smoke and industrial stacks was found to be the main pollutant killing the lichens, and, in many areas, the forests as well. The early chapters address this subject and another major industrial pollutant—fluorides—that caused problems wherever the soils or ores that contained them were processed.

As the USA prospered in the 1950s, the automobile population grew and with it the emissions from cars, trucks and buses, especially when trapped and concentrated in such air basins as Los Angeles. Already by 1946 this pollution had reached serious proportions. Research soon addressed photochemical air pollution, including ozone and oxides of nitrogen. After four decades this is still the major pollution concern, not only in the USA but also in all other industrialized as well as developing nations.

As a result of the early work on sulfur dioxide and fluoride, and subsequent study of these pollutants along with photochemical pollution, we now have a reasonable idea of what these pollutants do and how they do it. What we lack is a precise measure of the concentrations and exposure periods that cause a given effect. The present state of such knowledge is discussed as well as the interactions of the many climatic, soil and other potential abiotic and biotic stress factors. Interactions among pollutants, ranging from synergistic to mitigating, also are treated.

Innumerable 'trace' pollutants exist in the atmosphere. Some, such as pesticides, mercury and hydrogen sulfide, are important mostly in local situations and are not discussed in this volume. Others, including nitrogen-based air pollutants, are serious regionally and are treated. Other chemicals that occur both naturally and from human activity are of global concern. These include carbon dioxide, methane, fluorocarbons and stratospheric ozone. Because of the potential impact on plant health and distribution these also are discussed.

Biologically, there are many unknowns towards which research is directed. The much publicized decline and death of forests in Europe and the USA is but one. The reductions in crop yields and the resulting economic impact is another. We know that photochemical air pollution, largely ozone, is pervasive over much of the USA, eastern Canada and Europe, but we still have only a general idea of what effects ozone may be having on unmanaged, natural ecosystems, including rangelands and forests, which are vastly more complex than the monoculture of the farm.

The status of knowledge on these and other air pollution topics will be covered in this book. We shall look at the real and potential effects of air pollutants on the whole plant and on plant communities and explain these effects on the basis of fundamental biochemical and physiological mechanisms. The sources and dispersion of pollutants are both of interest and importance to understanding a problem in a given area so, although not a major thrust of this volume, these subjects will not be ignored.

In order to integrate the plant responses and weave a more meaningful story, the information will be presented from a historical perspective, very much as it came to happen and underscoring the underlying reasons for it. In this way we hope to bring together the most relevant air pollution literature in a manner that is comprehensive, authoritative and readable.

This book should serve as both text and reference for anyone concerned with how plants respond to air pollutants. It should serve the needs of anyone in agriculture, forestry or ecology who wants to know how air pollutants, either alone or interacting with the rest of the environment, might be influencing plants. The subject of air pollution biology overlaps a number of fields, including agronomy, biochemistry, ecology, entomology, forestry, horticulture, meteorology, plant pathology, plant physiology, soil science and many more. Further though, the book discusses the economic aspects of air pollution on plants and is intended to provide guidance to decision makers in the many regulating and research agencies, and others who share a concern for the quality of our air environment and the direct and indirect responses of plants to it. We hope that this volume will be as interesting and helpful to those who know nothing about air pollution as to those who have many of the answers.

It was tempting to omit reference notations in the text to facilitate reading. But in the final analysis their importance was felt to warrant inclusion where they were cited. However, to reduce redundancy, a single reference list at the end of the volume is provided rather than after each chapter. A list of suggested readings is, however, provided at the end of each chapter.

Warmest thanks are offered to those who helped with the manuscript; Richard Adams for outstanding constructive criticism on Chapter Fifteen, on the economics of pollution, and Dieter Deumling for a critical review of Chapter Twelve, concerning forest decline. Thanks also to Judy Baker for the prompt and accurate typing of the manuscript.

Michael Treshow
Franklin K. Anderson

CHAPTER 1

Introduction

WHEN IS AIR POLLUTED?

This book is about air pollution. More specifically, it is about how air pollution influences plant growth in agriculture, the forest and other ecosystems. First though, we should ask, 'What is air pollution?' 'What does it do?' The answers may seem simple, but when examined closely they are not. Air can be said to be polluted when its natural uses are impaired. When the sulfurously pungent taste of an acrid atmosphere burns your throat, there is no question. When the irritation leads to coughing or even choking, and your eyes burn, there is no question. Or when a thick mist blocks visibility, is there any doubt? But what about lethargy, depression or difficulty in breathing? The effects of pollution can be subtle for humans and plants alike. A gradual decline and ultimate death of whole forests can result from air pollution, but the slow insidious demise of forests may escape notice because it can take years. A loss in crop production is even more difficult to quantify. Still more subtle is a gradual change in concentrations of a natural component of the atmosphere such as carbon dioxide, whose increasing concentrations over a period of decades may adversely affect both the animal and plant life of the whole world.

Air is the invisible, tasteless, odorless mixture of gases that surrounds the earth. Some gases in this amalgam can be harmful in higher concentrations. Some components, such as oxides of nitrogen or sulfur, are always present, but excessive quantities can impair human health and agricultural production, and kill forests.

Such consequences may be the dramatic end result of air pollution, but most commonly the effects are less striking, at least in the short run. Every environmental factor influences plant growth, productivity and survival. Together with biotic stresses imposed by insects, fungi and viruses, the abiotic stresses related to climate and soil conditions also are critical. Similarly, the air environment is one more vital component of the total

1

Plant Stress from Air Pollution

environment on which the life of the planet depends. Alone or in combination with the other aspects of the environment, air quality is critical to plant and human health alike.

Excessive amounts of sulfur and other polluting chemicals come largely from human activities; thus they are called anthropogenic. There are obvious sources, such as smelters or coal-fired power plants, that emit sulfur dioxide (SO_2). There are also phosphate and aluminum reduction plants that emit fluorides, and a myriad of other sources that emit an enormous miscellany of wastes in lesser amounts. But most pervasive of all in modern society are pollutants from automobiles and other vehicles that contribute so much to human welfare, but also to particulate pollution and what we call photochemical pollution.

Many other contaminants of the air also can be found, but generally they are distributed rather locally. These can include various heavy metals such as lead and cadmium, or the metalloid arsenic. Mercury and chlorine also can occur in local pollutant episodes, as can ammonia, methane and many more pollutants. Together with a vast diversity of hydrocarbons and other organic compounds, they give to the atmosphere—or detract from—its 'character'.

The destructive effects of air pollution also include accelerated corrosion and wear, soiling of materials and reduced visibility. These deserve mention, although they are not the subject of this book. Moisture alone can reduce visibility to a few feet, as anyone knows who has experienced a dense fog. Whenever the relative humidity exceeds about 70%, the effect of moisture on visibility is pronounced and significant. Below this level, the influence of moisture is more obscure. Yet even in relatively dry air, when particulate materials or certain gases are present, their interaction becomes important. Moisture, however, influences the capacity of gaseous pollutants as well as particulate materials to impair visibility. Sulfur oxides in combination with water vapor, and the presence of nitrogen oxides, are especially effective in scattering light and thereby obscuring visibility. Hydrocarbons, oxidants and a multitude of exotic waste gases contribute further to urban, and now rural, haze as well as to still more harmful consequences. Air pollutants cannot only reduce visibility in terms of visual range, but they can alter the colors and contrasts perceived in the landscape when viewed through polluted air.

DISTRIBUTION OF POLLUTANTS

As a general rule, air pollution arises first from some local source. Pollution may originate on a scale as small as a backyard fire or, on a far larger scale, as from the burning of thousands of tons of coal each day in coal-fired electric power-generating plants. These are both examples of stationary, point sources. Or, pollution may originate from multiple sources, often mobile, as emitted by automobiles and other vehicles. When there is only one point source, it is relatively easy to trace the pollutants that are released from it and also any problems related to the health and welfare of residents in the area. When a multitude of sources are located over a given area, however, a single individual offending source is difficult to define. Such is the case in industrialized areas and urban centers in general, where the numerous pollutants both from stationary and mobile sources mix and disperse over large areas, polluting whole air sheds.

Air sheds can be thought of as air masses that are somewhat delimited by natural geographical boundaries, and that have properties that are relatively homogeneous. A classic example is the area around Los Angeles, California, where the San Bernadino Mountains to the north and east form a natural barrier that tends to confine most of any pollutants within it. This is known as the 'South Coast Air Basin'. In Europe, the Ruhr provides an important example; in England, the Trent Valley is another.

Pollutants produced in any air shed obviously are not completely confined, especially during periods of high pollution and air stagnation. As the lower air layers fill, the pollutants drift through the mountain passes and over the lower hills, as in the California example, or they can drift seaward to return to land farther up or down the coast. Or, polluted air over large populous areas of the eastern USA or Europe may be transported regionally with the general air flow. Thus, the contaminated air can, and does, disperse to areas well beyond the air shed of their origin, and it may cover half a continent.

It is not always easy to delimit an air shed or even a regionally contaminated air mass. In the eastern USA, and over much of northern Europe, the boundaries of air sheds are obscure. The contaminated air shed may extend hundreds of kilometers. When meteorological conditions favor stability, such as when a high-pressure system with its accompanying inversions covers a broad region, the trapped pollutants accumulate in the air mass, and harmful quantities may spread over vast areas. It is at such times that a polluted air mass may blanket an entire region, such as most of the northeastern USA, and persist for several days or even weeks. The polluted air mass will drift with the prevailing air movement during these days and may damage crops and forests hundreds of kilometers from the place of origin.

Theoretically, such contaminated air could become globally dispersed, as we see with the more publicized radioactive clouds of bomb tests, nuclear accidents and volcanic eruptions. But generally, dilution from broad dispersal, trapping of pollutants by larger particles, falling out in precipitation and diffusion into the ground and vegetation limit the spread. Ultimately, pollutant concentrations become reduced to non-harmful levels, often within the first hundred kilometers of their origins.

HISTORICAL PERSPECTIVES

Standing by a camp fire once, breathing in the irritating smoke that went with it, I thought of how much worse the smoke might have been in the caves and huts of earlier human populations. Smoke and air pollution must have been part of our environment since we first discovered fire. They were the trade-off we had to make for warmth, cooked food and safety from predacious animals and perhaps even other humans.

Indoor pollution, especially in developing nations, is still critical where wood, soft coal and dung provide the principal fuels for heating and cooking. Carbon monoxide and hydrocarbons in the smoke are more than irritating: they can impair health and shorten lives. Thus, for perhaps the majority of people in the world, such air pollution is a very significant part of life. Those who live in the relative affluence of modern society might seem to be less threatened.

The classic paper by John Evelyn, *Fumifugium, or the Inconvenience of the Aer and Smoake of London Dissipated*, written in 1661, portrayed the risks of coal 'smoake' to people and plants alike. This is a fascinating account of the times, and includes such observations that the smoke 'kills our bees and flowers abroad, suffering nothing in our Gardens to bud, display themselves, or ripen . . .' He wrote that the 'fuligenous and filthy vapor' made the city of London resemble 'the court of vulcan . . . or the suburbs of Hell'. He accused particularly the brewers, dyers, lime burners and soap boilers. Evelyn suggested that the smoke be controlled by using taller chimneys to carry it away, and the odor masked by the planting of odoriferous and fragrant flowers.

When Sam Clegg warned against the sickness and oppressive headaches resulting from illuminating gas in 1841, the public paid little heed. More widely notable pollution episodes of greater impact arose with the Industrial Revolution several decades later. Pollution then became a way of life for greater numbers of people than ever before. The industrialized society saw the burning of more coal in the smelting of ores, and production of iron added to the pollution from wood-smoke, tanning operations and decaying rubbish in the streets.

Hydrochloric acid associated with the development of the chemical industry was recognized as an air pollutant in the nineteenth century. It was produced in the making of sodium carbonate and did severe damage to property and vegetation. Hydrogen sulfide from tar distillation and nitrogen oxides from sulfuric acid manufacture also must be added to the list of early chemical pollutants. All this pollution of air in the city was an extension of the pollution of the air of the factory, so that industrial and urban pollution became one.

Sulfur was released into the atmosphere in vast quantities when ores high in sulfur were roasted to extract copper, zinc, lead or other elements. Once released from the ores and coal, it combined with the oxygen in the furnace and atmosphere to form sulfur dioxide. This gas was irritating to people but lethal to plants in the concentrations formed. Thus, early reports of air pollution impact appeared from England and Germany that described the decline and demise of forests and woodlands near the smelters. Similar impacts on agricultural crops caused even more concern. The injurious elements in the smoke were in doubt until Morren in 1866 and especially L. von Schroeder in Germany in 1872 proved that sulfurous acid was the important toxic agent responsible for much of this damage. Arsenic, lead, zinc, copper and other metallic emissions that were of concern were less toxic. The next few decades saw some significant scientific contributions on the subject in the German literature.

At the same time, across the channel in Wales, the Swansea Valley made its name in the annals of air pollution history. Here, an incredible 600 furnaces were smelting ores along the banks of the River Tawe. It was the center of the world's copper industry. The toxic smoke seemed to be the natural companion of prosperity, and little was mentioned in the literature of the day. The first copper ore mined in the USA was refined in Swansea, and it was late in the nineteenth century before the first smelters were built in the USA. The story of devastated forests in Tennessee and Montana is the story of research into understanding the effects of specific air pollutants and the development of pollution control technology. The time between recognizing a problem and doing something about it was measured in decades, and it was not until after 1910 that much appeared in the scientific literature. Prior to that period, 'air pollution' was referred to as 'smoke', much as today 'hazardous wastes' assumes essentially the same meaning as 'toxic wastes'.

Far more time elapsed between the first recognition of injury from fluoride and efforts to correct it. Accounts of fluoride 'toxicoses' in cattle can be traced back to the Icelandic literature of nearly a thousand years ago. These writings describe how domestic animals suffered dental afflictions, became sick and died after eating grass contaminated by pollution following volcanic eruptions.

It was much later that fluorides from industrial processes were found to be harmful. Again, the impact came with the Industrial Revolution late in the 1800s. Fluoride is abundant in certain minerals, and when these are heated it is released into the atmosphere. The manufacture of bricks and ceramics, the reduction of phosphate ores, which contain considerable fluorides, and the smelting of aluminum all produced fluoride pollution. With the recognition of fluoride effects in Germany in the late 1800s and with a newly emerging aluminum industry in Switzerland at the turn of the century, it is surprising that it took several years of aluminum production in California and Washington in the 1940s to recognize similar problems.

Returning to the urban front, smoke caused continued concern in the most polluted cities from the mid-1800s. However, decades of inactivity passed, other than for initiating some local ordinances, before meaningful legislative actions were taken. A number of studies were made of the smoke situation in major industrial cities in the decades preceding the First World War. Chicago conducted an intensive study in 1910 that resulted in a 500-page report involving soot, ash and smoke. St Louis had done the same in 1907, and Pittsburgh in 1912. Similar studies were begun in Great Britain at the same time. Most led to ordinances, but not much was accomplished by way of enforcement. One reason was the war and the subsequent post-war economically adverse situations. Economic depression produced higher priorities for survival. It was not until after the Second World War, in 1946, that affluence and priorities came together, and concerns were again seriously directed towards improving air quality (Haliday, 1961).

Additional air pollution episodes further motivated action. Elevated pollution combined with several days of still air led to disasters in such places as Meuse, Belgium, in 1930; Donora, Pennsylvania, in 1948; and especially, London, England, in 1952. These events aroused the public and stirred it to action. In Great Britain public alarm led to the formation of an Air Pollution Commission and to passage of the Clean Air Act of 1956 and to the demise of coal burning in the cozy but noxious fireplaces of urban England.

The air of London tasted more acrid than usual that December 5th in 1952. Throughout the day a still mist, more than a typical London fog, clung oppressively to the ground, forming a black shroud of sulfurous fog. For five days it grew denser, and by the end of this period 4000 deaths *more than normal* for a similar period had been reported. Many more individuals suffered increased respiratory tract symptoms: coughs, sore throats, vomiting and bronchitis. Chronic cardiac vascular disease also was prominent in the London episode and in similar incidents recorded in the air pollution literature.

This and other classic episodes were most common during the winter months, when atmospheric temperature inversions and air stagnation are most pervasive. Consequently damage to the vegetation, which is relatively dormant in winter, was not readily apparent. Later research was to demonstrate that plants can be damaged in the winter, despite their more dormant state.

Just as the effects of city smoke and industrial pollution were beginning to be recognized, and cities were coping with 'smoke' pollution, a new kind of air pollution arose in the Los Angeles area that provided further stimulus for controls. When it appeared the air became generally hazy and irritating throughout the city, and often well beyond it. The pollution could not be traced to any single industry or local source. It seemed to appear from nowhere. It was not smoke or caused by smoke. With the reduced visibility came sore throats, running nose and eyes, and headaches. Leaves of many agricultural crops in the area, especially lettuce, endive and spinach, became glazed and brown.

Air pollution experts came from the east to study the situation and decided that sulfur dioxide was responsible. But remedial measures installed by industries and oil refineries to curb sulfur dioxide largely were ineffective. The pollution prevailed despite control of sulfur dioxide. Further investigations, enhanced by new analytical technology, such as infrared spectrophotometry, opened an exciting new area for study: that of photochemical air pollution. The dirty, orange-brownish haze became known as 'smog', a term created decades earlier in England that alluded to a combination of smoke and fog. Although a technical misnomer, the word caught on and came to refer more broadly to almost any kind of air pollution, but especially the oxidizing atmosphere of modern cities.

This 'Log Angeles type' smog, or photochemical oxidant pollution, first recognized in the mid-1940s, eventually developed in the atmospheres of every affluent city in the world in the ensuing decades. Poorer cities, with their equal density of vehicles, also became affected. Research to understand this amalgam of atmospheric pollutants led to the development of many new methods of study, both chemical and biological.

Still more recently, another threat from air pollution was reported in air pollution literature, and, indeed, in the popular news media of the world. This problem quickly became known and popularized as 'acid rain'. It was first proposed as a threat to the forests of Sweden in the 1970s. By 1976 a major conference sponsored by the US Forest Service was held in Ohio. The proceedings were published in a 1074-page volume. During the next decade hundreds of volumes and thousands of articles were written on the subject in both scientific and popular literature. 'Acid rain' became the catchword of air pollution in the 1980s and a major new stimulus for research. A plethora of harmful effects on terrestrial as well as aquatic ecosystems was proposed. Many diverse roads were taken in the search for impacts that led to new approaches to discoveries, as well as to incredible political interplay.

Since the early days of air pollution study that roughly coincided with the Victorian era, great strides have been made in understanding the problems and nature of pollution effects. As with so many fields of investigation, there were several stages in the evolution of air pollution science as a biological discipline. Each successive stage became more sophisticated in response to new questions, new discoveries and progress in technology. Developments also reflected the thinking of the times and the social and political atmospheres. The courses of action were much the same in the USA and in Europe, with differences largely in the timing. There are in the world many developed countries still without significant air pollution control laws and technologies whose practices are similar to those of our own pre-war era. Developing nations, too, still have many pollution episodes ahead of them, and many lessons to learn.

SUGGESTED READING

Allaby, M. (1983). *A Dictionary of the Environment, 2nd Edn.* New York University Press, New York (Distributor: Columbia University Press, New York). vi + 529 pp.

Durham, J.L., ed. (1984). *Chemistry of Particles, Fogs and Rains.* Butterworths, Boston, Massachusetts. Acid Precipitation Series, vol. 2. An Ann Arbor Science Book. From a symposium at Las Vegas, Nevada, March 1982. xiv + 262 pp., illus.

Finlayson-Pitts, B.J. & Pitts Jr, J.N. (1986). *Atmospheric Chemistry: Fundamental and Experimental Techniques.* Wiley–Interscience, New York, xxx + 1098 pp., illus.

Jones, H.G. (1983). *Plants and Microclimate: A Quantitative Approach to Environmental Plant Physiology.* Cambridge University Press, New York. xviii + 323 pp., illus.

Manion, P.D. (1981). *Tree Disease Concepts.* Prentice Hall, New York. 399 pp.

Speidel, D.H. & Agnew, A.F. (1982). *The Natural Geochemistry of Our Environment.* Westview, Boulder, Colorado, Westview Special Studies in Natural Resources and Energy Management. xvi + 214 pp., illus.

Treshow, M., ed. (1984). *Air Pollution and Plant Life*, Wiley–Interscience, New York, Environmental Monographs and Symposia Series. xii + 486 pp., illus.

CHAPTER 2

Plants and Damage to Plants

Air pollution—several kinds of pollutants—can damage plants. This is taken for granted. The evidence is overwhelming. The damage is often dramatically visible in the form of white to colored markings on leaves and fruit, or burnt, dry necrotic lesions that eventually drop out of living plant tissues, leaving ragged-edged or shot-holed leaves. These are the obvious symptoms of severe, or acute, exposure by aggressive air pollutants—elevated levels of fluoride, sulfur dioxide, chlorine gas and others. These are symptoms that are seldom seen any more, but which once were common around large industrial centers—trees and whole forests denuded and killed by smelter smoke. This was a common occurrence at the turn of the century, through the 1950s in developed countries, and even today in developing nations.

Some of these pollutants were the by-products of smelting, sintering or manufacturing processes; others were the local result of accidents such as the breakdown of chlorination equipment at a swimming pool or emissions from a derailed and broken tank car.

But today these extreme occurrences are rare. They are no longer the sort of pollutant events in which researchers are interested. Today it is the near-background levels that attract attention. The long-term, low-level or chronic episodes of pollutant exposures are the ones mostly being studied today. The goal is to discover 'threshold levels' above which some adverse effect is measurable. One purpose of such studies is to guide in the enactment of realistic legislation that is strict and effective, but not *too* strict. Too much regulation is as bad as none at all. Over-regulated industries are unable to remain in business, which brings about adverse effects of another sort, perhaps just as serious.

The final word on the amount of any pollutant that causes harmful effects to plants or animals remains elusive. Such 'threshold' values are highly controversial. The reasons are complex. A major problem also exists in

9

that, even if we could agree on an air quality standard, the next step of attaining this standard, such as through the control of emissions, remains elusive. In fact, many cities that have high pollutant concentrations are designated as 'non-attainment' areas due to the relative inability to meet desired standards.

Two main problems exist: accounting for concentrations of pollutants that occur naturally, and attaining a desired air quality standard. For example, the total suspended particulates (TSP) can be high in remote, non-urban areas because of wind-blown dust. Ozone concentrations also can be high in remote areas, but how much is natural and how much due to long-range transport? Mechanisms of achieving photochemical pollution standards are especially elusive. The use of catalytic converters on automobiles and greater engine efficiency are helpful but costly. Also, there is still a question whether or not such approaches have improved air quality.

CHANGING AIR POLLUTION CONCERNS

Over the years, as our experience has grown, 'smoke' as a pollutant was replaced by specific, named pollutants. These were associated with specific industries or industrial processes. Then began a series of refinements in our thinking of air pollution. Not only must we be concerned about local pollution, but now we find it is necessary to examine questions of regional, national and even global pollution. As the world's population grows, these questions will become more serious. A tall stack was once considered to be a good method of reducing ground level pollution, but now this is recognized to aggravate regional pollution. In addition to the clearly labeled traditional pollutants, we now must be concerned with the precursors of secondary pollutants, with the dynamics of long-range transport and pollutant sinks, and with synergisms—the interrelationships of two or more pollutants acting at once.

New phenomena appear regularly on the pollution scene: simple smoke gave way in our thinking to the components of smoke, sulfur dioxide for example. This in turn has assumed new importance as the precursor of acid rain. Acid rain has now almost run its course as the popular or 'theme' pollutant; instead, ozone, both 'good' and 'bad' ozone, is becoming the new bandwagon pollutant. Good ozone is the stratospheric kind that shields us from harmful ultraviolet radiation. This kind of ozone is believed to be threatened by CFCs—chlorofluorocarbons (e.g. freons), which are widely used in aerosol cans, solvents and as refrigerants. Bad ozone is the ground-level kind that forms photochemically, especially in urban areas where it stings eyes, oxidizes materials and damages sensitive plant tissues. It is one of those odd and annoying realities of nature that CFCs are apparently not reactive with the bad ozone, only the good. Thus we see how the emphasis changes on different aspects of air pollution as scientific understanding, politics, media attention and environmental perceptions change.

We are now discovering that atmospheric carbon dioxide, once considered to be 'natural' and inoffensive, is rapidly rising in concentration, leading to alarm over a possible 'greenhouse effect' of rising temperatures on a worldwide scale. This in turn may trigger a host of alarming consequences, ranging from the melting of the ice-caps to the flooding of coastal cities by rising sea levels and impacts of CO_2 on modifying sensitivity to air pollutants. At the same time, the plant components of major ecosystems will be altered in response to the changing temperatures, and perhaps to the carbon dioxide concentrations in themselves.

On the horizon there are other pollutants that will become the new targets of public, political and professional concern. Perhaps these might include re-entrained pesticides from agricultural spraying, gaseous emanations from long-forgotten landfills, or hydrocarbons released naturally by vegetation.

Visibility through the atmosphere is affected by pollutants, both particulate and gaseous. This is becoming an important target of public concern and regulatory attention. By 1988 the US EPA had even begun to regulate the emissions from wood-burning stoves. New alarms will be raised, new pollutant effects will be discovered, and new studies will be done. As long as population rises, and also the gross national product, it is unlikely that air pollution will ever be 'controlled'; i.e. reduced to levels that have no significant adverse effects on some important systems such as forest trees, crops, esthetic values or human health.

FOCUSING ON SITES OF AIR POLLUTION DAMAGE

When pollutants at low, or slightly elevated, concentrations damage a plant, what is it that is damaged? One might say at first that the leaves are being marked, that there are spots or bands on them, or holes in them. But to mark a leaf in this manner is to mark a plant tissue or a cell, or a group of cells. So the questions remain, what is the damage, and what is damaged? Looking at a cell, one might say that the cell wall was damaged, or that some other part of the cell was attacked. Which part? How attacked? If some organelle of a cell were the target, what was the mechanism? If a chloroplast is bleached by sulfur dioxide, how? If a mitochrondrion is disrupted, or its function is somehow blocked, what has happened? The answers to these questions lead always to smaller and smaller levels of plant organization; e.g. from leaf tissues to cells to organelles, and finally to molecular levels. From organelles and membranes it is only a small step to the structures visible only through the electron microscope. These 'ultrastructures' in turn are made of molecules—proteins, lipids, carbohydrates—and even more fundamental chemical species than these. At length the building blocks of the plant substances themselves must be examined, the amino acids, which are constituents of plant proteins. Indeed, the very sequencing of the precursors of amino acids, the nucleic acids of DNA, may be involved. Even at the submolecular, atomic level where individual chemicals act and react, there is action affected by air pollution. It is the goal of several lines of research to discover the active mechanisms each type of pollutant has on plants. There are, of course, many possible sites of injury and many mechanisms of injury.

Photosynthesis

Photosynthesis in green plants is the most important chemical process on this planet and probably the most sensitive to air pollutants. It consists of the conversion of light energy to chemical energy. It is the manufacture of sugar from carbon dioxide and water in the presence of light, with oxygen produced as a by-product. Without photosynthesis there would be no animal life, no oxygen in our atmosphere, no fossil fuel reserves, and perhaps not even any free water in the environment because of the 'greenhouse effect'— the build-up of excessive heat trapped by an atmosphere that would otherwise consist largely of carbon dioxide. The few other autotrophic processes that exist in the earth, the chemosynthetic reactions in sulfur-metabolizing bacteria, for example, do not produce oxygen. Green plants can produce several liters of oxygen per hour for each gram of chlorophyll they contain. One kilogram of live leafy tissue contains about 1 g of chlorophyll. The cells of many algal species contain even more chlorophyll compared to other tissue weight: up to 4 g/kg. In addition to oxygen production, a large part of the chemical energy stored in all organic materials originated during photosynthesis. Our fossil fuels were originally formed from green plants that grew in warm climates, probably in the presence of much more abundant atmospheric carbon dioxide than is now available. These plants grew and increased at a higher rate than they were consumed by decay or other destructive agents, such as fire.

Photosynthesis is generally expressed by a balanced chemical formula similar to the following, either with or without the water on the right-hand side:

$$6CO_2 + 12H_2O \overset{light}{\rightarrow} C_6H_{12}O_6 + 6O_2 + 6H_2O$$

The water shown on the product side of this particular statement of photosynthesis is put there to emphasize that photosynthesis is now understood to proceed by two stages: a photochemical or light-dependent, energy-capturing reaction that produces oxygen, and an enzymatic light-independent stage that involves additional chemical reactions, including the production of some water, and which is often referred to as the 'dark reaction'.

All photosynthesis in higher plants takes place in chloroplasts, within membrane structures called lamellae, which in turn contain subunits called grana. A gel-like matrix, the stroma, surrounds the grana. Enzymes for the dark reactions are located in the stroma along with chloroplast DNA, ribosomes and some other substances. Each granum consists of even smaller, flattened sacs known as thylakoids. Adjacent grana are interconnected by the thylakoid membrane. It is in these highly organized substructures that chlorophylls and other pigments are found, and it is the thylakoid membrane that seems to be most vulnerable to air pollutants. Chlorophyll in solution, isolated from these structural entities, does not carry out photosynthesis.

In the photochemical or light reaction stage of photosynthesis, pigment molecules—chlorophyll a (and also 'accessory pigments', e.g. chlorophyll b and carotenoids)— capture light energy and use it to split water molecules, a process called photolysis. When water molecules are dismantled in this process, oxygen is produced, electrons are passed to electron receptors, and the hydrogen atoms go to the creation of $NADPH_2$ (reduced nicotine adenine dinucleotide phosphate). The electrons, which move through a series of electron receptors, give up some of their energy in steps along the way to produce energy-rich molecules of ATP (adenosine triphosphate). The ATP and $NADPH_2$ are now available and are utilized as energy sources in the dark reaction phase of photosynthesis to drive the reactions of carbohydrate synthesis, in which energy-rich carbon compounds are built up out of carbon dioxide. Some new water molecules are also produced in this process. These energy-rich carbon compounds, products of the dark reaction, include not only simple glucose but also sugar phosphates, which are the raw materials for the production of sucrose, starch, cellulose and other plant products.

The fact must be underscored that even the reactions of protein synthesis are intimately linked to photosynthetic processes. Carbon dioxide, water, nitrate ions, sulfate ions and other materials, e.g. various mineral elements, react in the presence of chlorophyll and light to produce oxygen and many organic compounds, including carbohydrates, lipids, amino acids, proteins, pigments and others. Nitrate, for example, is the source of nitrogen in plants, which is essential in the synthesis of amino acids, proteins, nucleotides, nucleic acids, coenzymes and chlorophyll itself. Sulfate is the form in which sulfur enters the plant, where it is essential to the production of some amino acids, some proteins and the coenzyme A critical to energy transfers. Nitrate and sulfate enter the plant along with other necessary elements, such as potassium, calcium, phosphorus, magnesium, iron, and others from the soil, in the water solution that enters the roots.

It may be disturbing to note that nitrate and sulfate, two of the substances accused of being precursors of acid rain, are here seen to be essential raw materials of photosynthesis. But this is so. Enzyme systems that act in protein synthesis in plants are intimately linked to the primary reactions of photosynthesis. Both nitrate and sulfate are such important nutrients to plants that without them amino acid and protein synthesis could not proceed. Atmospheric nitrogen (N_2), for example, is unavailable to plants. It must be provided in the form of soluble ions, either as nitrate (NO_3^-) or ammonium (NH_4^+)—but not too much!

It should be obvious to any thoughtful observer that in photosynthesis, and in the associated chemical synthesis processes, there are numerous steps that could be blocked or interrupted by outside agents such as air pollutants. Not only can the chemistry in the plant cell be disrupted, but the sites of chemical action themselves (cell and chloroplast membranes, grana, thylakoids, ribosomes, mitochondria, etc.) can be destroyed.

Plant organization

There are more than 200 000 different kinds of vascular plants, so no one species can be considered truly typical. The most familiar are the seed-producing gymnosperms and angiosperms. These include the coniferous and deciduous trees, shrubs, vines, grasses and herbs that are of greatest concern to agriculture. Their responses to air pollutants are economically and esthetically most critical, and understanding their responses requires some understanding of their structure as well as physiology.

Shoots and roots

The generalized plant structure consists of the shoot and root. The shoot is made up of the stem and leaves. The stem provides a structural framework for plant growth and gives photosynthetic tissues in leaves a favorable exposure to light. Through it, water and nutrients from the soil are transported throughout the plant, and organic molecules manufactured in the leaves are moved to the roots and other plant parts. The stem further stores food and gives rise to the reproductive structures.

Stem tissues are largely protected by the outer cell layers and provide little access for atmospheric gases to enter the interior. Hence, pollutants rarely affect these tissues. Roots, on the other hand, particularly in the pines, firs and spruces, are normally associated with mycorrhizal fungi that enhance the uptake of water and nutrients. Should air pollutants upset the normal carbohydrate balance in the roots, this association may be disrupted and plant growth adversely affected.

Nutrient and water uptake through the roots is generally not influenced directly by air pollutants since they are unlikely to enter the soil atmosphere in significant amounts. However, heavy metals or other toxic components in the soil can be harmful directly, or indirectly, when transported through the plant. These can be deposited on the soil by air pollution, or they may be made more mobile when present in soil minerals by changes in soil acidity caused by air pollution.

Leaf

The leaf is the principal photosynthetic organ of the plant and the organ most subject to the influence of air pollutants. When fully developed the tissues of the stem merge with those of the leaf. The leaf therefore has basically the same tissues as the stem; epidermis forms the outer layer, vascular tissue is arranged in veins, and photosynthetic tissue occupies the same region as does cortex tissue in the stem.

In the leaf, whether net-veined (dicotyledons) or parallel-veined (mono-cotyledons and conifers), the vascular tissues are so arranged that no part of the leaf is more than a few cells away from a vein. The veins consist of two cell tissue types: xylem and phloem. The xylem tissues are largely composed of tubular cells that begin to function passively to carry water from the roots to the leaves only after the xylem vessel cell is fully developed and dies. Phloem tissue is composed of living cells that have the capacity to move sugars and other products of photosynthesis to other parts of the plant by active transport.

The ground, or background, tissue in a leaf is called mesophyll. In addition to containing the vascular bundles, it is composed of two types of chlorophyll-bearing cells: elongated palisade cells that lie just beneath the upper epidermis, and loosely packed spongy mesophyll below the palisade layer. The center of each palisade cell is occupied by a large vacuole. The living cell contents, or cytoplasm, which is filled with chloroplasts, is located in a thin layer near the cell surface. The spongy mesophyll cells are also packed with chloroplasts but lack such a large central vacuole. Air spaces abound within the leaf, both within the palisade layer and especially within the spongy mesophyll. Each cell surface is bathed with water from the nearby vessels—water containing a dilute mixture of nutrients and minerals taken up by the roots from the soil solution.

The cell surface area exposed to air within a leaf is large. This humid internal atmosphere communicates with the atmosphere outside the leaf through tiny openings called stomata. Each stoma is defined by two lip-like guard cells that can swell as their internal water pressure changes, causing the space between them to open. Outside air containing carbon dioxide enters the leaf. Simultaneously oxygen can exit through these openings.

Stomata are part of the epidermal tissues of the leaf—a layer of flat, largely transparent cells that are often provided with a waxy cuticle. Stomata usually occur on lower leaf surfaces but also may occur on the upper surface and on most green stems in herbaceous plants. In woody plant stems the lenticels of the bark perform a similar function of gas exchange.

Stomata normally open in daylight and close at night, during drought or when excessively disturbed, as in a wind storm. But their behavior is complex; water relations (turgor pressure), extremes of light intensity and other factors such as carbon dioxide concentrations and the presence of air pollutants can affect them. Stomata are important factors in the sensitivity of a plant to air pollution because most of the gases entering or leaving a plant must pass through them. The epidermis, except at the stomata, is covered by a waxy layer that makes it fairly resistant to the passage of both liquids and gases.

THE DAMAGE PROCESS

It is difficult, if not too simplistic or even impossible, to generalize as to the precise mechanism by which each specific pollutant affects plants. The overall gross symptoms are distinctive for each pollutant. The precise way in which specific pollutants interfere with normal metabolic processes varies with the pollutant. Yet, a few commonalities are shared. These are summarized here only to provide a general background.

The epidermis is the first target of air pollution as the pollutant first passes through the stomata of the epidermal tissue and acts on this opening. Passing into the intercellular spaces, it dissolves in the surface water of the leaf cells, influencing cellular pH. It may next react with the walls of the mesophyll cells. Actually, the cellulose walls are relatively inert, and the pollutant is more likely to react with the cell membrane, notably the protein components. It does not necessarily react in its original form. Rather, passing into solution, the pollutant may form free radicals that, with their electric charges, may be still more reactive and toxic. Pollutant effects are not limited to the outer, cytoplasmic cell membrane, but may pass through this to come into contact with the cell's organelles, most readily the chloroplast. Here, it is the inner, thylakoid membrane that is apparently most sensitive. The enzymes of the thylakoid and protein components of the membranes are the most likely targets of pollution. The precise proteins affected would rationally vary with the pollutant, but enzymes essential to carbon dioxide fixation appear to be especially sensitive. Changes in the ultrastructure of the various organelles provide the first actual symptom of injury. These 'symptoms', like the foliar symptoms, vary with the pollutant. Chlorophyll cannot act in photosynthesis by itself without the membrane structure and ultrastructure provided by intact chloroplasts.

Specific reactions of the pollutant with an enzyme substrate are exemplified by sulfur dioxide, which oxidizes and breaks apart the sulfur bonds in critical enzymes of the membrane. The ratios of oxidized to reduced forms of sulfur are important to the integrity of the proteins, and the ratios are readily altered by the pollutant itself or the free radicals produced.

Although membranes appear to be most vulnerable, especially the thylakoid, which is most vital to photosynthesis, other structures and processes also can be affected. Consequently mitochondria—structures where respiration takes place—carbohydrate and lipid metabolism, and practically every other plant process all have been shown to be affected by air pollutants. The net result of this is an unhealthy plant, at first showing no outward sign of stress. Even before visible symptoms appear, however, the plant is weakened and its growth is impaired. Ultimately, the visible symptoms characterizing the presence of specific pollutants may appear, and death may ensue whether the targets are the well-known higher plants or the lesser-known lichens.

POLLUTANT UPTAKE

It is not so much the atmospheric concentration of a pollutant that should concern us but the amount that gets into the plant. Thus, pollutant uptake becomes critical. Uptake of any air pollutant from the atmosphere depends on more than ambient concentrations. The conductance through the stoma, regulating the passage of ambient air into the cells, is especially critical. Such movement depends on the concentration gradient between the ambient air and the sorptive sites within the leaf. Uptake, also referred to as flux, is a function of the chemical and physical properties along the gas-to-liquid diffusion pathway. Pollutant flow may be restricted by the physical structures of the leaf or scavenging by competing chemical reactions (Guderian *et al.*, 1985). Since these conditions may change during exposure, the ambient dose to which the plant is exposed does not necessarily reflect the actual cellular exposure.

The initial flux of gases to the leaf surface is controlled by the boundary layer resistance (i.e. the amount of a gas able to contact the surface) (Guderian *et al.*, 1985; Tingey & Taylor, 1982). This is a function of the leaf orientation and morphology, including epidermal characteristics, as well as air movement across the leaf. At slower wind speeds, less than 2 m/s, the boundary layer thickness decreases as wind speed increases (Hill, 1971). Thus, more pollutant would enter a leaf when there is some air movement. In chamber studies differences in air exchanges can account for much of the variation in results of such investigations.

Of the morphological components influencing uptake, pubescence is important in that the leaf hairs provide a major, yet relatively inert, area of impact (Bennett *et al.*, 1973). Cuticular waxes also are important in limiting uptake, even through a thin cuticle.

Stomatal resistance, on the other hand, is critical. Resistance or, looking at it conversely, the conductance, is determined by stomatal number, size, anatomical characteristics such as the degree to which they may be 'sunken' in the leaf, and the size of the stomatal aperture. When closed there is little or no uptake. Opening is regulated by internal carbon dioxide content, hydration of the guard cells, temperature, humidity, light, water availability and nutrient status, most notably potassium. It is the osmotic gradient produced by the potassium ions in the guard cells that regulates the guard cell turgor and opening of the stoma.

Despite the apparent significance of stomatal conductance, results of studies attempting to correlate this with injury have been inconsistent (Guderian *et al*, 1985). The genetic sensitivity of individual species and cultivars remains the overriding determinant of injury.

Movement of a pollutant in the liquid phase from the substomatal regions to the cellular sites of perturbation must also be considered to be part of uptake. A pollutant encounters many obstacles along this intercellular pathway. Scavenger reactions, such as with ascorbate, may absorb or neutralize a pollutant. Or, as with ozone, it may react to form other toxic substances—aldehydes, ketones or hydrogen peroxide (Tingey & Taylor, 1982). Also, free radicals may be increased indirectly or by decomposition in solution.

Finally, it is important to emphasize that it is the pollutant concentration within the leaf, more than the ambient concentration, that is most critical to plant health.

SUGGESTED READING

Jacobson, J.S., & Hill, A.C., eds (1970). *Recognition of Air Pollution Injury to Vegetation: A Pictorial Atlas*. Air Pollution Control Association, Pittsburgh, Pennsylvania. vii + 45 pp., illus.

Kozial, M.J., & Whatley, F.R., eds (1984) *Gaseous Air Pollutants and Plant Metabolism*. Butterworths, London, xi + 466 pp., illus.

LaCassse, N., & Treshow, M., eds (1976). *Diagnosing Vegetation Injury Caused by Air Pollution*. Environmental Protection Agency, Washington, DC. xxi + 139 pp., illus.

Treshow, M. (1970). *Environment and Plant Response*. McGraw-Hill, New York. xv + 422 pp., illus.

CHAPTER 3

Lichens: Sturdy Barometers

It might seem reasonable to assume that the earliest scientific studies of the biological effects of air pollutants would have concerned human health or the demise of forests near smelters. But such is not the case. The earliest studies, though by less than a decade, involved the response of lichens. While lacking the obvious economic significance of the more prominent vascular plants, lichens still play an important role in nature as well as playing a valuable role in air pollution biology and history.

Writing in the *Manchester Flora* in 1859, A. J. Grindon observed that the lichen flora was 'much lessened of late years through the cutting down of old woods and the influx of factory smoke . . .' In another pioneering paper from 1866, Finnish lichenologist W. Nylander made a list of the lichens growing on the trunks of trees in the Luxembourg Gardens in Paris. He noted that lichens here were poorly developed or sterile and wrote that lichens provided a measure of the purity of the air and constituted a kind of very sensitive instrument for measuring this quality. Three decades later, L'Abbé Hue could find no lichens at all in the same areas (reviewed in Anderson & Treshow, 1984).

By the early 1900s, the demise of lichens was described around many industrial cities of northern Europe, the USA and Canada (Richardson, 1975). Although perhaps not vital to the general public of the time, the effects of air pollutants on lichens provided a barometer for what was soon to happen to more prominent plants.

WHAT ARE LICHENS?

Among the many living things in the world that have been affected by air pollution are lichens, modest plant-like organisms that are reminiscent of mosses, liverworts, mold growths, or even of paint splashes or bird droppings in some settings. They are not 'plants' in the strict sense of the word, although most of them look like individual plants. They are composite

19

organisms made of two unrelated entities, a fungus and an alga, living together in one plant-like structure (Ahmadjian, 1967, 1974).

The combined lichen organism, plant-like but lacking roots, lives by photosynthesis. In sunlight and with adequate moisture, the algal partner, which contains chlorophyll, does the work of converting atmospheric carbon dioxide to sugar, the main food for the partnership. Lichens take nearly all of their moisture and nutrients from the atmosphere. The fungal partner provides the main structure of the lichen, including a place within its tissues where the algal cells reside, receive water, and are protected from ordinary environmental hazards. This partnership exemplifies one of the most successful examples of symbiosis known to occur in nature (Hale, 1961, 1967, 1969).

Many lichens grow in the same habitats as mosses and liverworts; i.e. in moist places, on the trunks and branches of trees, on fallen logs, and on the forest floor, mainly in regions of high rainfall. Lichens often resemble mosses and liverworts in appearance and in response to environmental conditions. But lichens also colonize difficult, even forbidding, habitats, such as rock faces or soils in deserts, or high mountains above the timber line, and they live in arctic regions where no other kind of vegetation can survive. Some lichens even penetrate the rocks themselves and live within them, gradually dissolving and crumbling them into the initial components of new soil. There are even lichens that grow on glass windows, bleached bones, grave markers, fence rails, painted barns, concrete walls, and other such unlikely places where ordinary plants cannot grow.

LICHEN SENSITIVITY TO AIR POLLUTION

Lichens are famous for their ability to withstand harsh living conditions. Besides being found in some of the world's most adverse habitats, experiments have been done that show remarkable toughness among lichens. Different lichen species have demonstrated abilities to resist damage, survive, measurably respire, and even to reproduce after being frozen at near absolute zero temperatures. Some have resisted desiccation, heat, mineral and nutrient stress, lack of light, excess light intensity, and even being held in a vacuum for six years.

But lichens are intolerant of atmospheric pollution (Ferry *et al.*, 1973; Hawksworth, 1973). Lichens do not thrive in cities or near polluting industries. This intolerance is dramatic. Lichens do not merely decline in the most heavily polluted regions; they disappear. A mere handful of species are able to survive in polluted areas. Although there are different opinions concerning the actual agent of lichen destruction in polluted air sheds, sulfur dioxide is almost certainly the main pollutant responsible for their decline and disappearance.

This one characteristic of lichens, their inability to withstand polluted air, has brought them to research prominence in recent years. This interest in lichen response to air pollution has not been due to any particular love for lichens by biologists. Rather, it has coincided with the general awakening of popular interest and concern for the environment itself that has been evident since the 1960s. Lichens have come to be regarded as pre-eminent bio-indicators of air pollution. This knowledge is not new; what is new is the present-day concern for the environment itself. This attitude has led to a new evaluation of the role of lichens in the assessment of environmental degradation.

Perhaps in time, as knowledge of and familiarity with them increases, we shall see the lichens themselves accorded respect as worthwhile components of the environment. At present, although lichens are known to have their ecological roles, they are scarcely regarded as important; certainly not economically important. The loss of lichen communities is an esthetic loss to most casual observers, not an economic loss—not that this should be so! When whole forests of timber or huge acreages of crops are affected by acid deposition, photochemical smog, or fumigation directly by industrial emissions, no one thinks much of the loss of the lichen populations in the same areas. The crops themselves long ago replaced the pristine habitats of many lichens, forests, grassland species and other organisms. There have been times in the fairly recent past when even wholesale destruction of woodland areas was little noted. Today, however, environmental impact on forests is alarming, both esthetically and economically—and perhaps absolutely; no one doubts our dependence on nature any more. Perhaps some day we shall be able to regard the more humble inhabitants of the environment with the same appreciation we give to the more prominent species.

There is one peculiar exception to the notice given to lichens. When they disappear from a region because of air pollution, they are usually said to have responded to extremely low concentrations of the pollutant. Lichens are nearly always put at the top of the species sensitivity lists. Their intolerance of air pollution is so well known today that they are automatically assumed to be more sensitive than other plant species. There is a story in this—one that requires a deeper acquaintance with lichens and a comparison of them with higher plants.

There is no doubt that they disappear from polluted environments. There is no doubt that they are 'pre-adapted' to injury by air pollution. There is no doubt that they are adversely affected in a given area when higher plants in the same area show no similar effects. But it is a mistake to attribute all these effects to long-term low levels of air pollution. Some reports in the scientific literature have done this for regions where even the forest trees have been killed by the same pollution episodes. The lichens in some of these studies were said to be sensitive to long-term, average, low levels of the pollution while the trees were not. Many tree species are known to respond to a relatively few short-term, high-level peaks of pollution exposure. But when air pollution studies involving lichens are reported, there has been a tendency to correlate lichen responses with long-term, low-level average pollutant concentrations. With higher plants, where chamber studies have been possible and threshold levels are known, it is recognized that short-term, relatively high-concentration doses are required to do the damage. We believe it is so with the lichens too—it is the peaks of air pollution concentrations that do the damage. The peaks may not have to be so high as with higher plants, but it is the peaks that do the damage nevertheless. Moreover, there are stages in the life cycles of lichens that are at more risk to air pollution than is true for higher plants, which do not share these life cycle stages.

Dramatic decline of lichens in some areas has been correlated with extremely low ambient air levels of sulfur dioxide, but with little attention paid to peak concentrations. One quantitative scale for the estimation of sulfur dioxide air pollution in England and Wales, using epiphytic lichens, showed that all epiphytic lichens were absent at mean annual air levels of 170 µg of sulfur dioxide per cubic meter of air (0.06 ppm, parts per million, volume-to-volume basis) (Ferry *et al.*, 1973). This was designated as 'zone 0'. Then a series of severely lichen-depleted 'struggle zones' (zones 1–5) were indicated for sulfur dioxide levels between about 60 and 170 µg/m^3 (0.02–0.06 ppm). According to this scale, effects were measurable in terms of lichen abundance and vigor even at concentrations as low as 30 µg/m^3 of sulfur dioxide (0.01 ppm).

It is difficult to assess the true impact of sulfur dioxide exposure from such scales, because they do not give any information about the number, intensity or duration of the high-level peaks of sulfur dioxide exposure; or other exposure dynamics. As a result, there is disagreement over the question of whether these admittedly sensitive lichen species are responding to the long-term, average, low levels of air pollution or to the relatively infrequent high concentrations that always occur during fumigations. In either case, the lichens really do disappear from cities and industrial zones as reported. Lichen species richness and total cover are reduced (Muir & McCune, 1988). It has been unfortunate, but such studies have seldom been accompanied by reliable continuous monitoring of the ambient sulfur dioxide levels. Until this is done, it is hardly possible to resolve the question. If the sulfur dioxide concentrations are wanted, it would seem to be more reliable and scientific to go out with good instrumentation and measure sulfur dioxide than to count lichens and somehow deduce the corresponding sulfur dioxide levels. But when the early studies were done, no such sulfur dioxide instruments were available. Today they are, but to maintain a continuous ambient monitoring network is expensive, certainly too expensive to justify just for the sake of protecting or even studying lichens.

Although sulfur dioxide is perhaps the most important air pollutant affecting lichens, others have been reported. Pollution by fluorides, for example, while not as common as sulfur dioxide pollution, can be just as devastating (Perkins *et al.*, 1980). The same decline in species diversity and abundance that is found around sulfur dioxide sources has been observed in the vicinity of fluoride sources. Fluoride sources include aluminum reduction plants and phosphate plants. In all of the cases so far examined the lichen decline has been shown to relate to fluoride accumulation. The critical threshold for effects can be variable, depending on the environment and lichen species, but tissue levels of about 20–80 ppm (dry-weight basis) seem to be sufficient to cause damage or death.

Ozone has been recognized since the 1960s as a serious air pollutant affecting higher plants. But only since 1977 has its impact on lichens been recognized (Nash & Sigal, 1980). Few studies of the effects of ozone and other oxidants on lichens exist, but levels of 500–800 ppb (parts per billion = parts per thousand million) for 3 hours seem to be sufficient to cause color changes in the lichen algal cells without visibly marking the lichen thallus itself. Studies of lichens in the San Bernardino Mountains next to Los Angeles have shown a 63% decline in numbers of species, from 91 to 34, between 1913 and 1980. The main air pollutants of this region are oxidants, principally ozone.

The prominence that sulfur dioxide and other gaseous pollutants are accorded in air pollution literature has tended to obscure the role of particulates as air pollutants. Particulates are carriers of trace elements, heavy metals such as lead, metalloids such as arsenic, and other toxic compounds. Numerous studies have demonstrated the ability of lichens to accumulate trace elements, including heavy metals. Lead can accumulate in lichens growing along highways; when the level reaches about 1000 ppm it becomes toxic. Cadmium has been shown to be toxic to lichens at tissue levels of 300–500 ppm. Lichens also have been shown to be able to accumulate extremely high levels of zinc, copper, nickel, cobalt, chromium and iron, sometimes ranging from 1000 to 90 000 ppm. This suggests that, for some elements, lichens are quite tolerant.

LICHEN BIOLOGY

In the discussion that follows, which is meant to acquaint the reader with lichens as living organisms, the structural and reproductive features that are discussed are important to the understanding of air pollution effects on lichens. It will be seen that almost every morphological, physiological and reproductive feature of most lichens is significant to their tolerance (or lack thereof) to air pollution. The responses of lichens to air pollution is the main unifying principle behind this entire chapter, not lichen appreciation, classification, identification, physiology, morphology, or even ecology. The following sections are meant to suggest reasons why lichens have been thought to be the most sensitive of all vegetative organisms to air pollution.

The symbiotic partners

Lichens, while appearing to be independent plants and individual organisms, are in reality associations of fungi and algae that live together in a mutually beneficial symbiotic relationship. Both partners benefit from the combined association, which takes on a new growth form unlike that of either partner alone (Smith, 1921; Hale, 1967).

The fungi of lichens, numbering as many as 15 000 to 20 000 species, are mostly members of the great fungus class Ascomycetes, the so called 'sac fungi' because the spores of their sexual stages are borne inside sac- or bat-shaped terminal cells called *asci* (singular, *ascus*). A few are members of the fungus class Basidiomycetes, the club fungi, that includes the familiar mushrooms, which are named from the terminal cells (basidia) that produce their sexual spores externally on four (usually) prongs. Both groups have vegetative cells in the form of long filaments—mycelia—that propagate through soil and other nutrient media like microscopic strands of gossamer, absorbing food through their cell walls. When conditions are right, as in the autumn after a summer of vegetative growth, and when rains produce an abundance of moisture, these filaments join together into mats and masses that eventually produce fruiting structures. The most familiar are the cup- or saucer-shaped discs of the Ascomycetes and the mushrooms of the Basidiomycetes. In a lichen thallus the fungus component is the dominant member of the association and the one that accounts for nearly all its important morphological and reproductive characteristics.

The algae of lichens are more limited in kinds than are the fungi. About 21 genera of green algae and 11 genera of blue-green algae participate in lichen formation (Hale, 1969). These algae may be filamentous or single-cell in growth form, but all are photosynthetic organisms that require water, atmospheric carbon dioxide and sunlight to produce their sugars, the food on which both the algae and the lichen fungi depend. Although the fungal component of most lichens is capable of sexual reproduction, the algal components have never been observed to reproduce sexually when in a lichenized condition.

Lichens, being composite organisms, are not really unique 'kinds' of living things in the same way that orchids or dandelions are genetically distinct entities. The relationships among lichens depend upon the fungus component alone. The species of lichens are treated taxonomically as species of fungi.

The algal component is more like a host to an invading parasite, and it is the invader that dominates. Although the algal partner in a lichen sometimes contributes some useful characters towards identifying a lichen, it does not participate in any genetic or reproductive way in the lichen life cycle. The fungus contributes most of the structural tissues of the lichen, is the only member of the association that reproduces sexually, and it is the only member that produces any reproductive structures.

Lichen structure and the classic growth forms

Although the classification of lichens is determined by the fungal part of the association, their ecology is governed in large measure by the growth forms and physiological responses of the combined organisms to the environment. The thallus (body) of a lichen is nothing like the independent organism of either the fungal component or the algal component. In the combined lichen there is a new tissue organization and a new growth form. Indeed, there are many hundreds of growth forms in lichens, But the basic growth form categories are relatively few, perhaps half a dozen. The older classification schemes for lichens were more or less erected upon these growth form categories. Such systems are now obsolete, but the old categories remain useful and even indispensable to the study of lichens today.

The three main growth form categories are termed *fruticose*, *foliose* and *crustose*. These terms describe lichens that are, respectively, shrub-like in structure, leaf-like and crust-like. The growth form terms are a direct reference to and description of the arrangement and distribution of tissues within the lichen thallus. These tissue arrangements have a great deal to do with the relative sensitivity of the various lichens to damage by air pollution.

Fruticose lichens

These are usually shrubby in appearance, or string-like, strap-shaped, or otherwise upright or pendent forms that have no clearly differentiated upper and lower surfaces as would a leaf from a higher plant. Instead, they have a more or less uniform exterior surface that surrounds either a hollow or filled interior. A cross-section of a fruticose lichen would be roughly circular, with a thin outer rind-like layer called the cortex. Immediately under the cortex is a layer of fungus-enmeshed algal cells called the algal layer. Interior to the algal layer is a central region of cottony, loosely intermeshed fungal filaments (hyphae) called the pith or medulla. Sometimes the central part of the medulla is hollow. The medulla appears to function as a storage organ and as a region where unique lichen substances, called lichen acids, are produced and deposited. The algal layer is the region of photosynthesis in the lichen, but the sugars produced there are rapidly transported to the medulla, where they are stored as mannitol. Metabolic activity within the lichen is greatest in the algal layer, and it is this area that is most severely damaged by exposure to air pollution.

Foliose lichens

Foliose lichens, possessing the same internal tissues as the fruticose lichens, have them organized into a flat, leaf-like structure with definite upper and lower surfaces. The upper surface contains the upper cortex and the algal layer. The lower surface contains the medulla and a new structure, the lower cortex, which may be quite variable in color and texture from the usually mineral gray-green of the upper cortex. The undersides of foliose lichens may be black, white, brown or other colors, and they may be supplied with a sparse to dense coating of hair-like processes called rhizines, tomentum or simply holdfasts. Although root-like in appearance, and even sometimes called rootlets, these structures do not function as organs of absorption but only as organs of attachment. Sometimes these hairs occur around the margins of a foliose lichen thallus or of one of the reproductive spore-producing structures called apothecia; they are then called cilia or fibrils. Other structures encountered on either surface of a foliose lichen include bare patches, holes, pores and depressions called variously cyphellae, pseudocyphellae or simply pores, depending on their size and the details of their structure. The lower surface is also sometimes covered with a series of anastomosing ridges and folds called veins.

In wet weather lichens absorb large quantities of water, passively and sponge-like, turn brighter green, and are more metabolically active. In dry weather they quickly lose their moisture to the atmosphere and become dull-colored, brittle, hard and metabolically sluggish.

But lichens have none of the structures common to many higher plants that serve to protect or isolate their internal tissues from adverse environmental conditions, such as a waxy cuticle on their leaves, closable pores (stomata) or a vascular system to replace lost moisture and nutrients. Indeed, lichens have no leaves. The interior of a lichen thallus is more or less continuously exposed to the outside environment at all times.

When episodes of air pollution occur, such as a downwind fumigation from a smelter or large city, the lichen becomes totally immersed in the toxic atmosphere. If this happens when metabolic activity is high, the algal cells can be severely damaged; the chloroplasts can be killed, and chlorotic and necrotic lesions can occur in both the algal cells themselves and the lichen as a whole. Even during the so-called dormant periods that occur during hot, dry conditions, atmospheric pollution can enter the lichen interior and cause damage. Moreover, chemical fallout, both nuclear and ordinary toxic chemical fallout, will pass into a lichen thallus without difficulty because of the exposed, unprotected lichen structure. Acid rain or deposition much more easily penetrates a lichen thallus than it does a waxy-cuticled leaf surface.

Not only do atmospheric chemicals and gases enter a lichen more easily than they do the tissues of higher plants, but also lichens do not have any structures or mechanisms for the removal of toxic substances. Leaves of higher plants can be dropped when damaged by chemicals, but lichens, lacking leaves, retain everything that enters them except for whatever might be washed out passively during a rainy period. When chemical levels build up in the leaf of a higher plant, a necrotic spot will appear in that region as the tissues there die from exposure, at the same time isolating and immobilizing the toxic substances that caused the condition. The whole plant may suffer; productivity will decline; radial growth in the wood of the trunk may decrease; crops will be reduced. But the lichen has no parts to shed, no regions of growth to be diminished, no fruits to be prematurely dropped or starved and reduced in size. When toxic levels build up in a lichen, the whole organism is at risk.

If toxic levels are insufficient to damage a mature lichen thallus, then the propagules—spores and/or tiny vegetative disseminules—can still be damaged. They are even more sensitive than the adult lichens because they are more exposed to the environment; the result may be the disappearance of whole lichen communities through the lack of replacement by reproduction. The structure of lichens is thus seen to be of fundamental importance to the responses of lichens to environmental stress, including air pollution exposure.

Crustose lichens

Crustose lichens are similar to foliose lichens in their tissue organization, but they grow so tightly appressed to their substrates (rock, tree bark, twigs, glass, cement, bare wood and other such places) that they cannot be lifted free without damaging them. Indeed, many crustose species are capable of actually dissolving rock and penetrating into the surface layers, where they grow within the rock itself. Many others live within the bark of trees instead of on their surfaces, exposing only the openings of their reproductive structures to disperse spores. Such lichens have no exposed lower cortex at all. Many of these lichens are among the most difficult to identify. They have fewer distinguishing characteristics than the fruticose and foliose forms. Many must be examined microscopically in order to find suitable identifying features. The crustose lichens include among their numbers some of the most brightly colored of all lichens. There are brilliant oranges, yellows, chartreuses, greens, grays, whites, browns and blacks, in addition to many neutral shades of gray-green and olive. They have been imaginatively described as resembling paint splashes, bird droppings, rock shadows, blood splashes, frosting and even precious gems. Some of these colorful lichens have worldwide distributions, especially at high latitudes and high elevations.

Squamulose lichens

There are some groups of lichens that are somewhat intermediate in structure between the foliose and crustose lichens. They are foliose in having a primary thallus composed of minute, flat, leaf- or scale-like flecks of tissue called squamules. Many of these resemble crustose lichens because they have no lower cortex but only exposed medulla tissue. But they are not permanently attached to the substrate as true crustose lichens are. Some of these even resemble, and are called, fruticose lichens because their spore-bearing surfaces are elevated on pointed, goblet-shaped or sparsely branched little fruiting structures called podetia (singular, podetium) that arise from a squamulose primary thallus. A similar condition exists among lichens with truly crustose primary thalli, for relationships among lichens do not follow the growth form categories presented here but cut across them in sometimes bewildering ways. Sometimes squamules are found growing along the axis or branches of a podetium or a fruticose thallus. They are then called phyllocladia, a term that suggests their leaf-like character in these species. The main lichens exhibiting squamules, podetia and phyllocladia are several genera centered around the large and important genus *Cladonia*; they are therefore called the cladoniform lichens.

Among the cladoniform lichens are many that are important sources of food for caribou, reindeer and a few other mammals of the northern regions. As such, they can become the agents of concentrating radioactive substances that arise from bomb testing and nuclear accidents such as the one that happened at Chernobyl, Russia, during 1986. Radioactive fallout is, of course, a chemical and a kind of air pollution. Lichens completely lack both mechanisms and structures for excluding radioactive substances from their tissues. Once taken up by the lichen, such chemicals tend to remain chemically bound along with other metabolic products and stored in the medulla. When animals eat these lichens, the radioactive nuclides they contain can be passed and concentrated still further into bone, milk, meat and vital organs of the organism.

Fruticose, foliose, crustose and squamulose lichens all share one common feature. Their tissues are arranged in more or less continuous layers. The algal cells, called the phycobionts, are found in a definite layer of tissue near the upper or outer cortex. In older classification schemes these growth forms were all grouped together in a category called the stratose lichens. No longer useful for classification purposes, the term stratose still describes quite accurately the tissue organization of perhaps 95% or more of all lichens. We still call these the lichens with stratified or heteromerous tissues. Within the stratose lichens, i.e. in nearly all lichens, the phycobionts may be algae of either the great class called the green algae, or they may be blue-greens, but the green predominate.

Other growth forms

It should be emphasized that lichen growth forms are not discrete and immutable body plans. They are more like convenient rallying points or regions along a continuous spectrum of diversity. They may be likened to the principal colors of the rainbow: there is a continuous spectrum of colors, but only a handful of color names are used to describe the visible spectrum. There are many intergrading forms within the stratose plan, and there are other types of lichen thalli that are not organized on the stratose plan. Three non-stratose lichen growth forms are the granular or granulose lichens, the filamentous lichens and the gelatinous lichens. Most of these are less numerous (in species) and are less abundant (in numbers of individuals) than the more common growth forms already described—with one or two important exceptions.

Filamentous lichens are few in number. Their morphology is highly influenced by the filamentous algae that participate in the lichen partnership. They look more like strands of algal and fungal tissue than organized tissue structures. So also with the granulose lichens. These resemble sprinklings of dusty pale blue-green powder at the bases of trees or on moist rocks in shady places.

One of the more important of these minor growth forms is the gelatinous lichen type. The algal symbionts in gelatinous lichens are all blue-green algae. These exercise a greater influence upon the overall growth form of the thallus than usually occurs within the stratified lichens. In gelatinous lichens the thallus is composed of a tangled mass of combined algal and fungal filaments.

The algae of gelatinous lichens are often of the genera *Nostoc* or *Anabaena*, which are capable of fixing atmospheric nitrogen. Nitrogen from this source can then become available to other plants.

Cryptogamic soil

In the southwestern deserts of the USA there is a gelatinous lichen, *Collema tenax*, that grows on desert soil, forming what is currently called 'cryptogamic soil'. This is soil that is covered by a thick surface growth of gelatinous lichens. It is crumbly and porous. The soil where these lichens grow appears black to the eye, and it is often rough and minutely hilly, like a tiny moonscape with 2-inch high mountain scarps. To the initiated, encountering a pristine tract of intact cryptogamic soil is almost a spiritual experience, like walking on holy ground. Indeed, the initiated will avoid walking on it at all. The lichens protect the soil surface from erosion during wind and rain storms. They fix atmospheric nitrogen and release it as nitrate to this otherwise impoverished soil. When disturbed, as from the stamping of cattle and sheep in overgrazed areas, or when run over by off-road vehicles, the protective capacity is damaged and lost. If this happens when the soil is dry, as it usually is, the brittle surface is pulverized and can be blown away in a gust of wind. This exposes the unprotected subsoil to often catastrophic erosion. Even a careless footstep can start the process that could later become a tragic blowout and erosion area.

The role of cryptogamic soil has only been appreciated in recent years. There are presently efforts under way within the Bureau of Land Management and other governmental agencies to set aside and protect some of our remaining areas of still-intact cryptogamic soil.

There is evidence that cryptogamic soil formation and renewed lichen growth can be reinstated. Experiments have shown that existing healthy cryptogamic soil can be suitably broken up (in a blender) and redistributed over exposed soils to establish new colonies. If protected from disturbance these starts may eventually grow together to form a new continuous protective film of gelatinous lichen tissue that will once again protect and enrich the soil.

Cryptogamic soil is, however, delicate. It cannot tolerate much physical disturbance. It is likewise probably subject to all the impacts that lichens can suffer from during excessive exposure to air pollution, although there have not yet been many studies done in this area. There has been a tendency to claim that our western deserts are less liable to damage from air pollution than are the moist woodlands and other more mesic habitats and croplands of the east, particularly with respect to the effects of sulfur dioxide deposition in the form of acid rain. And perhaps this is true in most aspects. Most of our western deserts are, after all, quite alkaline and provided with a high buffering capacity against acidification. But that was before the role of the gelatinous lichens in the formation of cryptogamic soils was realized and appreciated. Perhaps there is cause for additional concern in this area if these lichens are liable to be damaged by sulfur dioxide directly, or by any secondary products such as acid rain.

Lichen reproduction

Lichen reproduction, like lichen structure, predisposes lichens to impact by air pollution. Lichens have been found on old headstones in some very old English cemeteries, but not on newer headstones. Mature lichens have been found in apparently healthy condition, but young colonies in the same habitats are absent. Observations like these suggest the reproductive units are failing to establish new growth. If the mature thalli are prospering, why are not new colonies appearing in some habitats? Air pollution has received most of the blame, although other causes have been suggested. The discussion that follows is not meant to acquaint the reader so much with the reproductive aspects and cycles of lichens as to show where those cycles are at risk to air pollution impact. The mechanisms of air pollution injury in the reproductive stages are different from those in mature thalli. Part of the difference relates to the more direct exposure to the environment that propagules must endure. Another aspect relates to the destruction or modification of suitable substrates for the establishment of new lichen colonies—destruction caused by air pollution and other environmentally degrading processes that have accompanied human expansion, urbanization and industrialization.

Vegetative lichen reproduction

Many lichen species possess a number of unique morphological features that are probably functional mainly in vegetative reproduction. Two of the most visible, prominent and important of these are structures called isidia and soredia (singular, isidium, soredium).

Isidia are wart-like, knobby, elongate or even branched projections that grow from the upper and lower cortex and from the margins of some lichen thalli. Their presence or absence, shape, size, location, distribution and density are all important in lichen taxonomy. In vegetative reproduction they break off and are scattered about to later grow and establish new colonies if they come to rest in a suitable habitat. Each isidium carries within its tissues both the algal and the fungal portions of the thallus.

Soredia are fluffy little tangles of fungal hyphae enclosing a few algal cells. Resembling tiny balls of fluff, they erupt through pores or larger openings called soralia. They are then blown about by the wind or carried floating on trickles of rainwater to new habitats. Some are even carried about by ants. Eventually, if they come to rest in a suitable location, they can grow to become new lichen thalli just like the isidia. The presence or absence, location, size, color and density of soralia and soredia are important taxonomic characters in lichen classification.

In the presence of atmospheric sulfur dioxide or other air pollutants, it probably becomes increasingly difficult for lichens to colonize new surfaces, whether by isidia, soredia or by means of sexually produced spores. In addition, pollutants may affect the substrate, as by acidification. Also, algal cells within soredia are much more exposed to the atmosphere and its pollutants than are those of intact lichen thalli. Sulfur dioxide is known to attack the chloroplasts of lichen algal cells and interrupt photosynthetic processes. Such an effect could account for the failure of new lichens that reproduce by means of isidia and soredia to become established in urban and industrial areas.

Sexual reproduction in lichens

What is true of the algal cell within a soredium or isidium is surely true of the naked algal cells that lie on the surfaces of moist rocks, on tree bark and in other habitats. Spores produced by lichen sexual mechanisms must fall near such cells in order to capture and lichenize them. If they have been damaged by air pollution the chances for successful colony establishment are greatly reduced.

Sexual reproduction in lichens is not well known. Spores, called ascospores in the vast majority of all lichens, are produced. These have been demonstrated to germinate in culture, and the cultured colonies have been grown. But the typical growth form of the intact lichen is never produced by the isolated fungal partner growing alone. The process of lichenization, the invasion of free-living algal cells by lichen fungal mycelia, has been observed in culture too. But the extent to which this occurs in nature is not known.

Ascocarps, the spore-producing fruiting structures of both lichens and the independent fungi of lichen species, have been observed in culture too, but the ability to produce these structures is gradually lost as the culture ages. It seems almost as if lichenization is essential to the sexual processes of these fungi. Isolated algal cells grown in culture are capable of zoospore production, which is a sexual reproductive process found among the green algae, but this never happens in an intact lichen thallus. The fungal partner, however, fruits freely in lichen thalli.

Lichen taxonomy and identification

It is the goal of all good taxonomists to group their chosen organisms 'naturally', i.e. according to relationships. The relationships they seek are the product of sexual processes and imply actual transfer or sharing of gene pools among closely related groups—transfer through time and sharing of various gene complexes by the species within a genus and the genera within a family, and so on. Not all trees, shrubs or grasses are closely related, despite appearances. So it is with lichens. The whole constellation of lichens is an artificial assemblage. Fruticose, foliose and crustose lichens, like trees, shrubs and grasses, are not necessarily related within their groups. True relationships cut across the growth form categories in every direction.

Lichens are today being treated as species of fungi. The algal partner has no part in determining relationship. It is not even very useful in simple identification of lichens. The same lichen fungus can successfully lichenize several different algae in some species. The growth forms of lichens are, however, useful recognition features in the identification of a lichen, whatever its true relationships. So are the various morphological characteristics described above: size, color, presence or absence of vegetative structures such as hairs, pores, soralia, isidia and a host of others. Some groups, especially among the crustose lichens, can hardly be determined to species at all without the aid of a compound microscope and a thorough acquaintance with fungal anatomy and terminology. Some species differ from their cousins only by spore size or the presence of a subtle and elusive cell type in the hymenium (ascus-producing tissue where spore formation occurs). Determination of a species that is lacking fruiting structures is impossible in most crustose lichens. Popular literature is available for the identification of many foliose and fruticose lichens, but crustose lichens have not yet received adequate treatment in the form of identification handbooks. Professional lichenologists have their monographs and journals, but most of these sources of information are not available to the lay person or often even to the interested biologist.

These considerations are important with respect to lichens and air pollution because many air pollution impacts are species-specific. There are sensitive lichens and lichens more tolerant to sulfur dioxide. There are proponents of the concept that lichens can be used as bioassay organisms for determining the presence and extent of air pollution. This is not an unlikely possibility. Lichens do disappear in polluted areas. They will sometimes disappear when higher plants in the same region will only be damaged slightly or remain unharmed. Lichens have also been observed to restock areas formerly too polluted to allow lichen survival. The listing, counting, measuring and mapping of the struggling lichen survivors can reveal a great deal about the air pollution effects of the area. It should be cautioned, however, that many lichens are not easy to identify and that conclusions based on faulty identifications can be misleading.

Lichen ecology

Habitat groups

In addition to the growth form categories, lichens can also be grouped by habitat preference. Some species can occupy several different substrates, but most are restricted to one basic type. According to this concept, all soil-inhabiting lichens are termed *terricolous*; those that grow on bare rock surfaces are *saxicolous* or *epipetric*; and those that grow on the bark of trees are called *corticolous* or *epiphytic*. These three categories are the most widely used, but others exist too. For example, *epiphyllous* lichens grow on the leaves of plants. *Lignicolous* lichens grow on bare wood such as fence posts, rotting logs, boards or shingles. Some lichens, termed *omnicolous*, can occupy several habitats indiscriminately, ranging from those already mentioned to brick, asphalt, asbestos roofing tile, cement, paint, sea shells, bone and even glass. Several *epizoic* lichens have been described that live on the backs of male Galapagos tortoises or on the wing covers of certain large flightless weevils and other beetles in tropical areas. A very few lichens are completely aquatic; others exist only in deserts, where they obtain all their moisture directly from the air. Just as for some higher plants, there are also lichens that are restricted to the tropics, the tundra, to temperate regions or to particular types of rock substrate, such as calcareous rock (limestone), or to other restricted habitats. But for the most part lichens are widely distributed and are found in even more habitats than many other plant forms can occupy.

All lichens respond strongly to the micro-environmental conditions that occur in their habitats, even more than to the larger mesoscale and seasonal climatic patterns they are immersed in. Within the limits of their environmental requirements, as with higher plants, lichen communities develop by slow stages through a sequence of successional, intermediate states to a sort of climax condition. The vertical distribution of lichens on a tree trunk, for example, is very pronounced because of the great variation that can occur from top to bottom of the tree owing to such environmental factors as light intensity, moisture, relative humidity, and exposure to wind, rain or pollution. This microclimatic gradient on a tree exists even though the entire tree and its forest of neighbors is immersed in the same long-term weather regime. This sensitivity of lichens to microclimatic variation is important with respect to air pollution. It underscores the need for research into lichen ecology in both healthy and polluted environments (Ahmadjian, 1974).

Tree lichens

Trees are often richly inhabited by lichen flora. As a tree grows, new regions of the trunk and branches are gradually colonized by a succession of lichens that first begin to appear on about the third internode from the growing tips (Denison, 1973). Foliose lichens appear first; crustose lichens next, on older internodes; and finally, on about 10-year-old internodes, fruticose lichens appear and begin to dominate. Eventually each internode ages and is overtopped with younger tree growth. The environmental conditions change. The overall effect of these changes on the lichens is dramatic. Each significantly different microhabitat is commonly occupied by a different lichen association. As the tree grows, the lower branches become shaded and sometimes die, at which time the process of change comes to a stop. The lichen community on the lower branches becomes a stable climax stage. In such an area the total biomass of lichens can amount to 505 kg/ha (450 pounds per acre), fresh weight. Some of these lichens whose phycobionts are blue-green algae can fix atmospheric nitrogen; amounts up to 11 kg/ha (10 pounds per acre) per year have been measured. This is an impressive example of one of the roles of lichens in nature. The possibility of harmful impact of high levels of air pollution or other disturbance, such as clearing, on such an ecosystem is obvious. The importance of investigating the lichen communities of undisturbed forests and discovering their responses to air pollution is also obvious.

Rock communities

In high mountains, in deserts, on lava flows, and in other places rich in rocks of large size and relative stability, lichens abound. They colonize the surfaces of these rocks in large patches, often round and reminiscent of gray-green mold growing on bread. A close examination of some of these colonies will reveal the inner regions of the larger colonies eroded and falling away. The lichen grows on at the edges in a roughly circular fashion, like a mushroom fairy ring. In a healthy unpolluted environment, the bare central areas of these large lichen patches are often colonized by young individuals of the same or a different lichen species. This is called community recycling. In regions that have become polluted through emissions from some large industry or an upwind urban area, recycling often disappears. The mature lichen thalli continue to thrive but no new lichens appear in their centers. The same phenomenon has been observed on grave markers especially in old cemeteries. The older headstones may be host to numerous large lichen colonies, but after a certain date no new colonies appear. The date may correspond to the appearance of a new industry in the vicinity. When the polluting industry is closed down, or effective controls are installed, recycling may reappear. These observations are clear indications that the reproductive units of the lichens, whether vegetative or sexual, have been more seriously affected than the mature, already established individuals.

Habitat destruction

It should not be assumed that all examples of lichen disappearance or decline are the products of air pollution. Lichens are probably more seriously affected by habitat destruction than by air pollution. Whole forests have been cleared for logging purposes or for agriculture, not to mention areas that have fallen under urban sprawl. These sources of lichen decline are seldom mentioned, while at the same time and in some of the same regions, the remaining lichens are said to be in trouble because of air pollution. It is an undisputed fact that air pollution, particularly sulfur dioxide, adversely affects lichens, and they disappear with the forest. But air pollution is not the only cause of lichen decline and disappearance. It is important to keep this fact in perspective (Ferry *et al.*, 1973).

SUGGESTED READING

Ahmadjian, V., & Hale, M. (1973). *The Lichens*. Academic Press, London. xiv + 697 pp.

Ahmadjian, V., & Paracer, S. (1986). *Symbiosis: An Introduction to Biological Associations*. Published for Clark University by University Press of New England, Hanover, NH. xii + 212 pp., illus.

Brown, D.H., ed. (1985). *Lichen Physiology and Cell Biology*. Plenum, New York. From a conference in Bristol, England, April 1984. xii + 362 pp., illus.

Ferry, B.M., Baddeley, M.S., & Hawksworth, D.L. (1973). *Air Pollution and Lichens*. Athlone Press, London, viii + 389 pp., illus.

Kershaw, K.A. (1985). *Cambridge Studies in Ecology*, Cambridge University Press, New York. x + 293 pp., illus.

Lawrey, J.D. (1984). *Biology of Lichenized Fungi*. Prageger, New York. x + 408 pp., illus.

Seaward, M.R.D., ed. (1977). *Lichen Ecology*. Academic Press, London. x + 550 pp., illus.

CHAPTER 4

The Early Days

The general decline and death of forests was the same, yet it was different. Over 100 years ago, forests began to die around the heavy industries that were then appearing in Europe (Stockhardt, 1853). All lacked any means of pollution control. Acid smoke spewed copiously from the vents and short stacks, or roasting heaps, of smelters, foundries, steel mills, acid plants and chemical manufactures. Smoke was presumed to cause forest decline and crop damage. But what were the toxic chemicals in the smoke?

Beginning some 100 years later in the 1970s, the forests of Europe again began to die. Why? The sulfurous smoke had been controlled long ago in many areas, and the forests had since recovered. Now forests far removed from industrial areas began a long, slow decline. The circumstances had changed, but the consequences were much the same. Plant pathologists and others had studied sick plants in the interim and had learned a great deal, but the explanations for the new decline remained elusive. In the early days the problems were even more mysterious since botanical knowledge was but a fraction of that which was to be learned over the next century. Plant pathology was barely emerging as a science, and the very nature of disease remained obscure and controversial. Fungi were often found to be associated with some diseases, such as the one killing the Larch forests of Germany in the 1860s. But were fungi the cause of the disease or the consequence? Disease was the will of God. The parasitic nature of disease was established only after the middle of the nineteenth century with the research on late blight of potatoes primarily in Ireland, and the rust fungi in Germany. Thanks to research by such men as the Rev. M. S. Berkeley and Anton DeBary, fungi became recognized as important parasitic pathogens of agricultural crops and forest trees. At the turn of the century bacteria also were recognized to be significant pathogens, but it was some decades later before the parasitic nature of viruses and mycoplasmas was discovered.

When parasitic organisms were found to cause disease, there was a natural tendency to blame every disorder on these and search for a causal organism. But it was not long before pathologists and foresters looked also towards the non-biotic environment for the causes of disease. Perhaps physical environmental stresses associated with the climate, the soil or the atmosphere could cause disease alone, without the involvement of some parasite.

The Victorian age of the nineteenth century, as portrayed eloquently by Dickens, is often viewed as a period of romance and elegance; of carriages drawn through parks, and serene landscapes. The fetid stench of open sewers, the burning of garbage, the fumes of industry small and large, more rarely portrayed, escape the images of the era. The gentle street-lamps burning coke gases and sending rays of diffuse light into a misty London fog display a romantic vision unfettered by the toxic air pollutants that accompanied it.

The general public was more concerned with survival in an era of high unemployment than the smoke from lighting, heating or industry. The foresters and plant pathologists of the day expressed concern, noting their fear that air pollution might damage plants. Compared with the canker of larch and other diseases caused by fungi, pollution was of minor concern. The *Handbuch der Pflanzenkrankheiten*, authored by Paul Sorauer and published in 1872, gave the subject scant attention. The same is true for the 1874 volume *Wichtige Krankheiten der Waldbäume*, written by Robert Hartig, widely acknowledged as the 'father' of forest pathology. In the English translation of this work, *Textbook of the Diseases of Trees*, Marshall Ward had two pages dealing with effects of 'coal smoke and the smoke from ironworks', the same space given to 'the effects of lightning'.

The earliest illness observed on deciduous trees, and later attributed to air pollution, was the discoloration of young leaves that ultimately turned mostly yellow and brown and finally died. In more severe instances leaves simply shrivelled and died. If one observed more closely, one would see that the yellowing was most prominent between the larger veins. This was accompanied by a reddening of the affected tissues on some species. The sudden expression of symptoms was considered as 'acute' injury by Sorauer (1914). This was distinguished from the 'slow poisoning', which he termed 'chronic' injury, that involved a more general loss of the chlorophyll. When plants appeared normal but production may nevertheless have been adversely affected, the term 'invisible injury' from smoke was used.

In both broad-leaved trees and conifers the symptoms first appeared in the tree tops. Leaves in the top gradually lost their color and died. The smaller limbs in the top died next, followed by larger branches, proceeding downward with the ultimate demise of the entire tree.

The early investigators noticed considerable differences in sensitivity to smoke injury among plant species. In Europe, birch and beech were the most sensitive broad-leaved trees, and spruce the most sensitive conifer. In the USA, white pine, hemlock, pitch pine and chestnut oak were among the most sensitive (von Schrenk and Spaulding, 1909).

The toxic nature of the smoke, laying day and night as a dense shroud permeating the vegetation, and its principal component, sulfurous acid, had only recently been discovered, but not without considerable controversy. Heavy metals, and metalloids such as arsenic, were considered by many to be the principal toxic component of the smoke. But it was sulfurous acid, or sulfur dioxide, that was largely responsible for the vegetation damage (Stöckhardt, 1871; von Schroeder, 1983; von Schroeder & Ruess, 1983). The previous views that damage was due to metallic poisons, including arsenic, zinc and lead, present in such smoke or soot were proved to be incorrect.

The earliest credit for establishing the toxic nature of sulfur dioxide goes to Morren in 1866 in the article 'Recherches experimentales pour determiner l'influence de certains gaz industriels, specialement du gaz acid sulfureux, sur la vegetation' that appeared in the *Report of the International Horticultural Exhibition, London*. M. Fretag (1869) established that the damage was not incurred through the medium of the roots. Toxic concentrations of sulfur were thought to be restricted to the immediate proximity of an industrial establishment such as a smelter or coke oven, or in a surrounding narrow valley where smoke lay dense throughout the day and night. Oxidation of the sulfurous acid into sulfuric acid when in contact with moisture was said to give some protection against the most extreme action of the poison. Smoke from lime kilns was considered less injurious because the lime retained the sulfurous acid.

The effects of acid mists on pine and fir were studied by H. Wislicenus and F. Schroeder in Germany in 1900 and in London in 1891 by Frank Oliver (*J. Royal Host Soc. London*, 1891). The effects of the mists were totally different from those of gases. Gases were only harmful when they entered the leaf. Fluids cannot enter the stomata, so liquids were considered harmful only when they evaporated. Sprays were injurious only at high concentrations, when they caused defoliation of the oldest needles and mild tip yellowing. But liquids did not cause the leaf burn or reddening caused by gases.

The Industrial Revolution, and the accompanying pollution, was initially a product of Europe. But the problem soon extended to the USA, and forests in both the eastern and western USA were extensively damaged by 1900. The chief of the US Bureau of Chemistry, J. K. Haywood, was among the first to study and write on the subject. As with so many subsequent studies, the early research was conducted in response to lawsuits. Haywood's work (1905) was undertaken at the request of the US Department of Justice in response to a suit brought by the USA against the Mountain Copper Company operating a smelter in a narrow valley in northern California. The basic question was: 'Did fumes from this plant injure vegetation, and if so, how large an area was affected?' Effects were measured in the amount of sulfur in the tissues, as well as visible symptoms. The episode was a significant landmark, although dead and dying pine and oak trees were observed miles further south. In Happy Valley large numbers of fruit trees were badly injured. Leaves of peach trees turned red and yellow and fell from the branches early in the summer.

The copper ores roasted in open furnaces were high in sulfur, which was given off to the air principally as sulfur dioxide. Inasmuch as the California ores contained just over 40% sulfur, substantial amounts of sulfur dioxide were liberated. Haywood calculated this as 748 tons each day.

The manner in which Haywood exposed young trees to known amounts of sulfur dioxide reflected the technology of the day. Plants were placed in a glass-sided cabinet 2 feet square and 4 feet high. A known weight of carbon disulfide in alcoholic solution held in a platinum dish was then set in the chamber. The measured amount of carbon disulfide released a calculated amount of sulfur dioxide into the atmosphere when burned. An electric fan in the chamber dispersed the sulfur dioxide uniformly. Typically, the plants were exposed to this gas for one hour. In this manner, the early researchers determined that 'as little as' 1 ppm sulfur dioxide was toxic.

Haywood (1905, 1910) went on to explore the role sulfur dioxide played in causing the decline of forests in Montana. Immediately surrounding the smelter practically all of the trees were either dead or severely injured. A few unaffected trees attested to the genetic tolerance of the survivors. Terrain and elevations varied considerably near this Anaconda smelter, and with this so did the plant species. The few ponderosa pine trees at nearby, lower elevations were most severely damaged, followed by the more prevalent Douglas fir trees. Lodgepole pine and subalpine fir showed symptoms at distances up to 30 km. The situation here was complicated by the high arsenic in the ores, but the particulate nature of this material limited the range over which it was dispersed. Accumulation of arsenic in nearby forage, and the losses of livestock feeding on it, are another story.

When ores high in copper were first discovered in the eastern USA in the 1850s, they were smelted in the Swansea Valley in Wales. By the 1890s copper producers chose to process the rich Tennessee ores locally. This may have been sound economically, but not biologically. Within five years an area of 7000 acres (2834 ha) became completely bare, and an additional 17 000 acres (6883 ha) was converted to grassland made up predominantly of broom sedge (*Agropyron scoparius*). A transition zone, consisting partly of more tolerant trees, comprised another 30 000 acres (12 146 ha) (Hursh, 1948; Hedgecock, 1914).

One more historic incident deserves to be mentioned for two reasons. It involved pollution generated in one country that caused significant losses in another country. Second, the case provided a tremendous stimulus in research and enrichment of knowledge of the effects of sulfur dioxide on vegetation. The lead and copper smelters were located in Trail, British

Columbia, and the damage extended well into the US State of Washington, 18 miles to the south. An international commission was formed to determine the cause and scope of the problem. Typical of many large smelters, emissions reached nearly 600 tons of sulfur dioxide per day. One advent of technology that began to appear during this period was the use of tall smoke stacks, mostly over 180 m, to disperse the pollutants. This may have lessened the risk to nearby vegetation but it greatly expanded the region impacted (Committee on Trail Smelter Smoke, 1939).

Smoke from industry was, in many cases, an extension of that from the city where coal had been used for home heating since at least the fourteenth century. No wonder that air pollution came to mean the same for most people as smoke, and perhaps sulfur dioxide. Much has been written about the effects of this smoke on humans in terms of public health and nuisance in general. But surprisingly little was written regarding effects on plants. It was only when these same sulfurous pollutants flowed from the stacks of smelters that concern was expressed and the results of studies appeared in the literature.

Research generated by the Trail, British Columbia, incident, as well as other episodes in California, Utah and Germany, first addressed the problem of which plants were most sensitive to pollutants. Next, emphasis shifted to determine the concentrations that caused symptoms. Even today the answer to this question is elusive. Sophisticated instrumentation now available still does not disclose the exposure regimes that are most critical.

The big question was what amount of visibly apparent leaf yellowing or burning caused a given loss in production. Was there a direct relation between injury and yield? Most researchers thought there was.

However, in 1923, J. Stoklasa in Germany proposed that metabolic processes might be suppressed, and yield reduced even when no visible symptoms were apparent. He called this 'invisible' injury since you could not 'see' such physiological changes. The semantics of this term led to unfortunate disputes. It was subsequently argued that this was a poor choice of words since a yield loss could be measured and thus the effects were not 'invisible' or 'hidden'. The term 'latent' has also been used. But it denotes a potential effect, not an effect in progress. 'Subliminal' might be a better choice of words; it pertains to effects that operate below a 'threshold of consciousness'. In any case, actual measurements of physiological changes had to await the development of more sophisticated analytical equipment.

Such instrumentation, though at first primitive by modern standards, followed in the 1930s. The early Thomas autometer, developed in the laboratory of the American Smelting and Refining Company in Salt Lake City by Dr Moyer Thomas, was among a few. It measured changes in conductivity of the sulfuric acid solution produced by the sulfur dioxide, or more accurately the sulfurous anhydride, absorbed in acidified water. It was sensitive to a number of interfering gases in the air, but was modified over the years to reduce this problem. Further progress came after refinements in infrared spectrophotometry. Discoveries that led to accuracy and reliability in air monitoring equipment were also critical, with the greatest advances made only after 1960.

Ozone monitoring has come a long way from its beginnings in the 1940s and 1950s, when the extent to which ozone caused rubber to crack was used as a measure of ozone concentration. The automated measurement of ozone oxidation of potassium iodide was the method of the 1960s. Chemiluminescent techniques followed in the 1970s, and ultraviolet spectroanalysis has been preferred in the 1980s.

Such monitoring equipment made it possible to dispense accurately known amounts of the pollutant being studied and thereby more critically evaluate the plant's response. However, it was necessary to grow plants in a relatively confined space, or chamber, into which the pollutant could be introduced and the concentrations monitored. The necessary improvements in chamber design followed, although the sophistication of today's open-top chambers is not vastly superior to the glasshouses in which sulfur dioxide effects were studied by Dr Moyer Thomas and his colleagues in the 1930s.

The melding of these technologies has made it possible to detect the most minute responses to pollutants at background or higher or lower concentrations, metered over any desired exposure periods, and the exposure dynamics computerized and programmed to the investigator's preference. Unfortunately, the most realistic and appropriate exposure concentrations and durations to use are still a matter of debate and study.

Such technologies were not applied for solely scientific reasons. The need for accurate response information was more fundamentally based on the need to justify control of the pollutants to meet air quality standards and to quantify the amount of control needed to protect forests, agriculture and human health.

The motivation for protecting vegetation was only partly altruistic. Lawsuits by growers to recover damages provided further incentive. This, coupled with public concern, and even public pressure in other instances, had two major effects. One was in the development of emission standards; the second was in advances and improvements in control technology, making it possible to meet the new air quality standards. Beginning in the 1960s in the USA, goals were directed towards setting air quality standards that would protect the public health and welfare. Welfare meant agriculture, natural ecosystems, visibility and materials. The concept of air quality standards is now widely accepted and utilized in much of the world. In the USA, air quality standards are constantly being modified in response to new knowledge as mandated by Congress. In Europe, many countries also have air quality regulations as well as emission standards, although many questions remain unanswered.

SUGGESTED READING

National Research Council of Canada (1939). *Effect of Sulphur dioxide on Vegetation.* NRC Publ. No. 815. Ottawa, Canada. 447 pp., illus.

Scheffer, T.C., & Hedgcock, G.G. (1955). Injury to northwestern forest trees by sulphur dioxide from smelters. *Tech Bull* No. 1117. USDA For. Serv. 49 pp.

Sorauer, P. (1914). *Non-parasitic Disease. Manual of Plant Diseases* Vol. 1. Record Press. Wilkes Barre, Pennsylvania.

CHAPTER 5

Effects of Sulfur Dioxide and Heavy Metals

Sulfur dioxide! This classic, pre-eminent air pollutant is still with us. Produced from burning coal, sulfur dioxide persists not so much from home heating as in the furnaces of coal-fired, electric power-generating plants. Particulate emissions may be over 99% controlled, and sulfur dioxide may be 70–90% controlled in some countries, but older plants or newer facilities in developing countries often lack such technology. Even where controls exist, the tremendous amounts of coal burned still result in the emission of significant quantities of sulfur dioxide. And despite pollution control equipment in modern smelters, the sulfur dioxide and heavy metal particulate emissions can still damage neighboring vegetation. The problem is especially critical in developing countries where control technology is lacking, or in developed countries where control has a low priority.

The frequent elimination or reduction of particulate materials from emissions has both good and bad aspects. The benefits are obvious, but particulates also serve to absorb some of the gaseous pollutants, including some sulfur dioxide. Thus, the captured (i.e. impacted) sulfur dioxide would settle out with the particulates close to a source. Without the particulates, sulfur dioxide remains as a gas and is free to travel far greater distances. The resulting lower concentrations of sulfur dioxide may not cause as much acute injury, but subliminal effects may be more prevalent and widespread than ever.

INJURY SYMPTOMS

Visible symptoms of sulfur dioxide injury are no longer pervasive, but where controls are inadequate or break down, as they do on occasion, or when unusual meteorological conditions prevail that concentrate the emissions,

44

plant damage can occur. The rapid breakdown of chlorophyll in the leaf cells results in characteristic chlorotic or necrotic markings. The basic cellular responses are similar for most species, but because of variations in their anatomy differences in specific symptoms are found on vegetable crops, cereals and coniferous plants.

Descriptions of symptoms can never equal the value of field examinations or color photographs in identifying or recognizing the cause of a given symptom. Excellent color illustrations appear in such references as the *Air Pollution Control Association Pictorial Atlas* (Jacobson & Hill, 1970) and the *Canadian Forest Service Information Report* NOR-X-228 (Malhotra & Blauel, 1980). A trilingual color plate atlas from Germany also is excellent (van Haut & Stratmann, 1970).

Vegetable and broad-leaved species

Some of the most sulfur dioxide-sensitive plants found occur in this group. The range of symptoms on dicotyledonous species is well illustrated by the response of alfalfa (*Medicago sativa* L.). Alfalfa is one of the most sensitive and important species affected by sulfur dioxide. Markings most often consist of white or straw-colored necrotic lesions extending through the leaf.

The mildest acute symptoms, which can follow one exposure of 0.3–0.5 ppm (300–500 ppb)* for a few hours, consist of a narrow border of necrotic tissue along the margins or tip of the leaflet. Necrosis may be sharply delimited and often extends irregularly between the veins towards the midrib (Figure 5.1). At higher concentrations, or during conditions of high light intensity and humidity, necrosis is largely intercostal. Affected areas may be of any size or shape, ranging from minute flecks to involvement of most of the leaf area. When chlorosis occurs it ranges from an almost normal green to pale yellow in color. When the chlorophyll is only partially destroyed the cell may remain functional.

The response of vegetable crops is similar. Squash and tomato are among the most sensitive species, and the ivory to white marginal and intercostal lesions are highly distinctive (Figure 5.2).

Cereal crops

Wheat and barley, notably the 'flag' leaves, are particularly sensitive to sulfur dioxide. Following 1–2 hours of exposure to concentrations as low as 0.3 ppm, leaf tips become gray-green to chlorotic; finally, they characteristically turn ivory-colored. Interveinal flecks or lesions often develop.

---

* The term ppb (parts per billion) is used throughout this volume to mean parts per thousand million (i.e. 10^9) as is the practice in the USA; ppm (parts per million) is commonly used with reference to sulfur dioxide and will be used with this pollutant.

Figure 5.1 Acute marginal and intercostal necrosis on alfalfa caused by sulfur dioxide

Woody perennials

Broad-leaved shrubs and trees may respond to sulfur dioxide with straw-colored interveinal lesions, but in many instances lesions may become dark or reddish-brown, or even reddish-purple in color.

Needle-leaved species

Conifers also may be highly sensitive to sulfur dioxide. The most pronounced symptoms of injury on such sensitive conifers as pine consist of bleached to reddish-brown necrosis beginning at the tips of the current-year needles (Figure 5.3). Older leaves, having lower metabolic activity, are most likely to be spared. Damaged needles may drop after a year or two rather than persist for the normal four or more years. Occasionally only a portion of

Figure 5.2 Marginal and intercostal necrosis on tomato caused by elevated sulfur dioxide concentrations

the needle is affected, producing a banded appearance. Less intense injury produces chronic chlorotic lesions. Necrosis and chlorosis often develop on scattered needles in a cluster rather than on all the needles.

The response during the winter to exposures of sulfur dioxide may differ. Injury is more apt to consist of a general chlorosis followed by a definite yellowish-brown banding. Gradually, over a period of several weeks, the affected interveinal areas fade to yellow, brown or reddish-brown.

PLANT SENSITIVITY

The concentrations of sulfur dioxide that are injurious depend on many factors. Genetic sensitivity, metabolic activity, the time of year, environmental conditions and exposure dynamics all influence plant response. Plants are most sensitive in the spring, when leaves are developing; but they may be injured even during the winter months.

There are two ways in which plant sensitivity can be evaluated. The easiest, although most empirical, is to observe plants near various sources of sulfur dioxide and estimate the relative amounts of visible injury (Jones *et al.*, 1979). This has the obvious disadvantage that the exposure dynamics that caused the injury are unknown. The second way to evaluate sensitivity

Figure 5.3 Characteristic needle tip necrosis of eastern white pine. Plants were
fumigated at 0.25 ppm for 2 hours

is to expose plants to known concentrations of sulfur dioxide for specific
periods of time.

The first, or field, approach to developing a ranking of comparative plant
sensitivity has provided some valuable insights. Such observations were
made around smelters in Europe in the 1800s. One of the earliest such
studies in North America involved observations in the forest around Trail,
British Columbia, beginning in the 1920s (Canadian National Research
Council, 1939). Innumerable further studies in the USA and Europe have
provided sensitivity lists for a number of diverse plant communities.

Each sensitivity list tends largely to represent species of a single community
or group of vegetative types in a geographical area. One such list that is
helpful to the diagnosis of plants in montane forest areas, such as are
common to the northwestern USA, is shown in Table 5.1. This list is based
on the Canadian National Research Council Report (1939), which provides
results and illustrations of extensive studies of sulfur dioxide effects. The

Table 5.1. Relative sensitivity of montane plants to sulphur dioxide[1]

Most sensitive	Less sensitive
Coniferous species[2]	
Larch	Englemann spruce *Picea engelmanii*
Larix occidentalis	Western white pine *Pinus monticola*
Douglas fir	Hemlock *Tsuga heterophylla*
Pseudotsuga menziesii	Lodgepole pine *Pinus contorta*
Ponderosa pine	Grand fir *Abies grandis*
Pinus ponderosa	White fir *Abies concolor*
	Red cedar *Thuja plicata*
Broad-leaved trees	
Paper Birch	Mountain maple *Acer glabrum*
Betula papyrifera	Red hawthorn *Crataegus columbiana*
Bitter cherry	Cottonwood *Populus trichocarpa*
Prunus emarginata	Mountain ash *Sorbus scopulina*
Aspen	Alder *Alnus tenuifolia*
Populus tremuloides	Choke-cherry *Prunus virginiana*
Shrubs	
Ninebark	Mountain laurel *Ceanothus sanguineus*
Physocarpus malvacearum	Oregon Grape *Mahonia repens*
Ocean spray	Raspberry *Rubus idaeus*
Holodiscus discolor	Currant *Ribes* sp.
Serviceberry	Snowberry *Symphoricarpos*
Amelanchier alnifolia	Thimbleberry *Rubus parviflorus*
	Elderberry *Sambucus coerulea*
	Dogwood *Cornus stolonifera*
	Sumac *Rhus glabra*

[1] Based on the National Research Council of Canada Report (1939).
[2] It is the younger needles that are highly sensitive. Mature and past-years needles are more tolerant.

sensitive species still provide the most valuable 'indicator' plants to use for field surveys.

The more quantitative, fumigation, approach has also provided excellent sensitivity lists. A major early study was conducted in the 1920s by P. J. O'Gara at the American Smelting and Refining Company Agricultural Station in Salt Lake City (Table 5.2). O'Gara examined the responses of over 300 plant species and varieties grown and fumigated under comparable conditions (Thomas *et al.*, 1950). The O'Gara factor that was developed represents the relative sensitivity compared with alfalfa, which was the most sensitive species studied.

There may be plant species occurring in local regions that are even more sensitive. For instance, the wild petunia (*Ruellia strepens*) appears to be more sensitive than alfalfa but occurs naturally only in a few counties in Missouri and Kansas.

Table 5.2. Plant species sensitive to sulfur dioxide[1]

Crop plants

Alfalfa	Cotton	Soybean
Medicago sativa	*Gossypium* sp.	*Glycine max*
Barley	Oats	Wheat
Hordeum vulgare	*Avena sativa*	*Triticum* sp.
Bean, field	Rye	
Phaseolus sp.	*Secale cereale*	
Clover	Safflower	
Melilotus trifolium sp.	*Carthamus tinctorius*	

Garden flowers

Aster	Four o'clock	Verbena
Aster bigelovii	*Mirabilis jalapa*	*Verbena canadensis*
Bachelor's button	Morning glory	Violet
Centaurea cyanus	*Ipomoea purpurea*	*Viola* sp.
Cosmos	Sweet pea	Zinnia
Cosmos bipinnatus	*Lathryus odoratus*	*Zinnia elegans*

Trees

Apple	Larch	Pine, Eastern white
Malus sp.	*Larix* sp.	*Pinus strobus*
Birch	Mulberry	Pine, ponderosa
Betula sp.	*Morus microphylla*	*Pinus ponderosa*
Catalpa	Pear	Poplar
Catalpa speciosa	*Pyrus communis*	*Populus* sp.

Garden plants

Bean	Lettuce	Spinach
Phaseolus vulgaris	*Lactuca sativa*	*Spinacea oleracea*
Beet, table	Okra	Squash
Beta vulgaris	*Hibiscus esculentus*	*Cucurbita maxima*
Broccoli	Pepper (bell, chili)	Sweet potato
Brassica oleracea	*Capsicum frutescens*	*Ipomoea batatas*
Brussel sprouts	Pumpkin	Swiss chard
Brassica oleracea	*Cucurbita pepo*	*Beta vulgaris*
Carrot	Radish	Turnip
Daucus carota	*Raphanus sativus*	*Brassica rapa*
Endive	Rhubarb	
Cichorium endivia	*Rheum rhaponticum*	

Weeds

Bindweed	Fleabane	Ragweed
Convolvulus arvensis	*Erigeron canadensis*	*Ambrosia artemisii-*
Buckwheat	Lettuce, prickly	*folia*
Fagopyrum sagittatum	*Lactuca scariola*	Sunflower
Careless weed	Mallow	*Helianthus*
Amaranthus palmeri	*Malva parviflora*	Velvet-weed
Curly dock	Plantain	*Gaura parviflora*
Rumex crispus	*Plantago major*	Poison ivy
		Rhus toxicodendron

[1] Plant sensitivity is based on presence of visible symptoms that make them useful as indicator plants in field studies.

PHYSIOLOGICAL AND BIOCHEMICAL MECHANISMS

Stomatal mechanisms

The waxy cuticle covering the epidermal cells of the plant provides an effective barrier to penetration by sulfur dioxide and most other air pollutants, oxides of nitrogen being a possible exception (Wellburn, 1988). Thus, for sulfur dioxide, the stomatal pores are the main entry ports to the internal air spaces. Accordingly, sulfur dioxide entry occurs mostly during daylight hours, under conditions of adequate moisture, when the surrounding guard cells are turgid and the stomata are open.

When the relative humidity is above about 40%, low concentrations of sulfur dioxide have the initial effect on cells of increasing the turgidity of the guard cells. This causes the stomata to open even wider and allows entry of still more sulfur dioxide, but at the same time enabling greater water loss in transpiration. A presumed mechanism for this action is that sulfur dioxide alters the pH, and the higher levels of bisulfite and sulfite disturb the fluxes mostly of potassium, chlorine and calcium between guard cells and the surrounding cells that regulate the stomatal aperature. Carbon dioxide is also suspected of having a role in the mechanism. Higher concentrations of sulfur dioxide, at least above 2 ppm, encourage stomatal closure. Sulfur dioxide-induced ethylene production may be involved in this process, but the evidence is not conclusive (Black, 1984).

Biochemical transformations

Once within the substomatal air spaces of the leaf, sulfur dioxide comes into contact with cell walls of the mesophyll cells. The cell wall is not especially reactive but a number of other reactions may occur. Sulfur dioxide readily dissoves in the intercellular water to form mainly sulfite (SO_3^{--}) and bisulfite (HSO_3^-) ions. Very little unionized sulfur dioxide remains in solution. Sulfite and bisulfite have a lone pair of electrons on the sulfur atom that strongly favor reactions with electron-deficient sites in other molecules. Sulfoxyl radicals (SO_2) also can be found that may be reactive in donating electrons to electron-deficient sites as well as allowing another means of oxidizing sulfite to sulfate.

Defense mechanisms also may come into play in response to low concentrations of sulfur dioxide. Oxidation of sulfite to the less toxic sulfate can occur by both enzymatic and non-enzyme mechanisms (Malhotra & Khan, 1984). Peroxidase, cytochrome oxidase and ferredoxin NADP reductase, and catalysts such as metals and ultraviolet light all provide effective mechanisms (Hallgren, 1978). Free radicals can be formed but can be impaired by protective natural scavengers. Ascorbic acid may be the most important scavenger. Excess sulfur dioxide can be transported from old leaves to young leaves, from leaves to roots, and from roots to the surrounding medium.

The release of hydrogen sulfide (H_2S) by plants exposed to sulfur dioxide appears to provide a major defense. Hydrogen sulfide concentrations may increase 1000-fold. A cycle of cysteine being desulfhydrated to sulfide and then oxidized to sulfate, then reduced to sulfide again, may enable the plant to expel the excess sulfur (Filner *et al.*, 1984).

Photosynthesis and energy relations

Before reaching the chloroplast, the various sulfur dioxide products must first pass through the cell wall and membranes. Although the wall presents a minimal obstacle the plasmalemma well may be disrupted in the process, leading to secondary changes that may affect the transport of ions. Transport is probably across the embedded protein components of the membrane, and adverse reactions may occur here. But the real damage occurs initially in the chloroplast membrane and within the chloroplast.

Ultrastructural studies show that swelling of the spaces within the thylakoids occurs (Wellburn *et al.*, 1972). In some species the grana may be reduced in size or disappear completely (Soikkeli, 1981). The shape of the chloroplast may undergo gradual changes, from ellipsoidal to oval to spherical. Invaginations, i.e. intrusions, from the inner membrane of the chloroplast membrane also become evident (Godzik & Sassen, 1974). Granulation of stroma occurs at later stages of injury. Increases in the size and number of plastoglobuli also are evident as early responses to sulfur dioxide exposure (Soikkeli, 1981).

Membrane and chloroplast disruptions are reflected in the decreased net photosynthesis by concentrations of sulfur dioxide as low as 0.035 ppm (Black & Unsworth, 1979). The reason for such sensitivity may lie in competitive inhibition by SO_3^{--} with HCO_3^- at the carbon dioxide binding site of the enzyme ribulose 1,5-bisphosphate carboxylase (RUBISCO). But the inhibition may not be competitive with respect to HCO_3^-. However, RUBISCO requires a pH change in the stroma from pH 7.5 to pH 9.0 before becoming active. Hence, is the response due to blocking of the reactive site, or to an inability to alter the pH adequately?

Sulfite also affects phosphoenolpyruvate (PEP) carboxylase involved in the C_4 pathway of photosynthesis. Again, competitive inhibition may be involved (Ziegler, 1973). Rapid binding of sulfite with PEP carboxylase would limit its availability for carboxylation. Bisulfite compounds have a similar inhibiting reaction (Mukerji & Yang, 1974).

Sulfur dioxide also can affect photosynthesis by attacking the photosynthetic electron transport and photophosphorylation reactions. Membrane damage can cause disorganization of the two photosystems since both are localized in the thylakoid membranes of the chloroplasts. Photosystem II appears to be most sensitive, and sulfur dioxide appears to inhibit the accumulation of the primary electron acceptor (Shimazaki & Sugahara, 1980). Integrity of the thylakoid membrane is critical to maintaining a proton gradient across it. In the event of damage to the membrane, insufficient ATP would be produced to sustain normal metabolic activity and growth.

When sulfur dioxide enters the leaf it is metabolized to sulfite, bisulfite and sulfate. These compounds in turn may affect several biochemical and cellular processes that can inhibit photosynthesis. Bisulfite is either directly incorporated into the sulfuric groups of the sulfolipids or is taken up at the binding sites in the thylakoids. Dithiol groups produced in the chloroplast as a result of photosynthetic electron transport provide a source of reductant for changes (Alscher, 1984). Enzymes on the thylakoid membranes are apparently especially effective in providing sites capable of picking up free sulfur dioxide and are thus an important site of pollutant activity.

Environmental factors are major determinants in the way in which sulfur dioxide may influence photosynthesis. Light, relative humidity, temperature and carbon dioxide concentrations may alter the flux of pollutant into a plant largely through their action on stomatal conductance. However, respiration or detoxification rates also may be affected; or environmental factors may directly affect the plant sensitivity (Black, 1984).

One mechanism of action for high sulfur dioxide concentrations is in causing the degradation of chlorophylls to phaeophytins, leading to early senescence (LeBlanc & Rao, 1975). This results in visible symptoms and is not likely to account for more subtle influences.

Effects of sulfur dioxide on enzymes comprise earlier influences. Sulfur anions inhibit ATP formation in mitochondria, and the supply of ATP is reversibly diminished as sulfur dioxide concentrations are increased, thus limiting the energy available for initial photosynthesis (Harvey & Legge, 1979).

Sulfur dioxide has been variously reported to inhibit, stimulate, or have little or no effect on dark respiration (Black, 1984). Conflicting reports are attributed in part to different methods employed to study respiration as well as the wide range of concentrations employed.

Photorespiration is an inefficient form of the dark reactions of photosynthesis in which oxygen is fixed instead of carbon dioxide; no carbohydrates are generated. Energy is used to metabolize phosphoglycolate, which results in a net loss of fixed carbon. The few studies with sulfur dioxide on photorespiration are conflicting. Increases of photorespiration with increasing sulfur dioxide concentrations have been shown that may result from the greater use of energy in repair or replacement processes (Kozial & Jordan, 1978). On the other hand, sulfur dioxide may drastically reduce or totally arrest rates of photorespiration (Ziegler, 1975). Since such inhibition would enhance net photosynthesis, it would be beneficial to the plant.

Lipids and other cell constituents

Glycerolipids constitute about 50% by weight of thylakoid membranes. Sulfur dioxide causes a pronounced reduction in their concentration and composition (Malhotra & Khan, 1984). Structural alterations in chloroplasts may be related to this. Decreases in lipid content may be brought about by either reduced synthesis, increased lipase activity, peroxidation of fatty acid chains or a combination of these.

Exposure to sulfur dioxide causes an increase in soluble sugars, possibly due to a breakdown of polysaccharides rich in reducing sugars. Sulfur dioxide also affects a number of enzymes and metabolites involved in organic acid metabolism, including maintenance of cellular pH.

The ultimate significance of the effects on carbohydrates may lie in translocation and storage. Phloem loading may be reduced, leading to less storage in the roots, which could have serious consequences to plant health and production. On the other hand, a number of studies have failed to show any yield reductions of vegetable crops in the absence of visible injury.

PRODUCTION AND GROWTH EFFECTS

Crops

For many years the amount of yield reduction caused by sulfur dioxide was considered to be directly related to the amount of visible leaf injury. This concept was questioned for decades (Stoklasa, 1923), and we now know that significant losses in production and growth can occur without another symptom ever being seen. Such 'hidden' or subliminal effects are well documented (e.g. Malhotra & Khan, 1984).

Three main approaches to the study of sulfur dioxide responses have been used:

(1) Observations and measurements of plants near point sources of sulfur dioxide.
(2) Fumigation studies in greenhouses or chambers, sometimes with controlled environment conditions.
(3) Fumigations in open fields with known sulfur dioxide concentrations but utilizing existing environmental conditions.

The sulfur dioxide concentrations known to cause visible symptoms range upwards from a minimum 'threshold' dose of 0.3–0.5 ppm for more than a

1–3-hour period. A detailed study based on 6500 field inspections by H. C. Jones and his colleagues at TVA indicated that a threshold dose for foliar injury of sensitive species was 0.7 ppm for 1 hour or 0.1 ppm for 8 hours, although the most sensitive species under certain conditions may be injured at slightly lower concentrations.

This represents some oversimplification, however. Uniform sulfur dioxide concentrations are rare. Exposures range from brief periods (i.e. peaks) of high concentrations to prolonged periods of lower concentrations. The peaks are most harmful. The frequency and duration of the exposure are also significant because interruptions between exposure periods allow time for some physiological recovery (Zahn, 1970).

These variables, plus the numerous environmental parameters that influence plant sensitivity, are taken into consideration in establishing guidelines or air quality standards for a pollutant. The current US standard of 0.5 ppm for 3 hours is under study and may be a bit lax. The guidelines provided by the World Health Organization are much lower. The responses of forests as well as crop plants are considered. Guideline values are given as: annual average, 30 μg/m^3 (0.012 ppm); 24-hour average, 100 μg/m^3 (0.038 ppm).

In one particularly realistic study (Guderian & Stratmann, 1968), 21 agricultural crops grown in large containers were placed at six sites at increasing distances from an iron ore roasting plant emitting sulfur dioxide. Mean sulfur dioxide concentrations were expressed as mean concentrations during 'exposure' periods when sulfur dioxide concentrations exceeded 0.10 ppm (260 μg/m^3), and as mean concentrations over the entire period. Visible leaf injury ranged from none to 25% leaf burning. Average sulfur dioxide concentrations ranged from 0 to 0.141 ppm (370 μg/m^3). Production was given as percentage of control, the control plants being those furthest from the source. The results (Table 5.3) showed that yields of sensitive crops were impaired when average concentrations of sulfur dioxide exceeded 0.01–0.02 ppm (about 26–52 μg/m^3). Unfortunately the critical peak concentrations were not given. Inasmuch as visible symptoms were generally present, the results reveal little about subliminal effects.

Occurrence of subliminal growth reductions in the absence of visible symptoms has been most clearly demonstrated with grasses. Following up reports of poor pasture production in polluted hillside regions of east Lancashire, England, rye grass was grown in glasshouses in Manchester and exposed to either ambient or filtered air (Bleasdale, 1973). There were consistently significant reductions in dry weight of plants grown in ambient air even when concentrations remained below 200 μg/m^3 (about 0.076 ppm) and no visible injury symptoms appeared. Subsequently, other grass species were subjected to still lower concentrations of sulfur dioxide (Crittenden & Read, 1979). Dry weights of shoots of *Lolium perenne* cv S23, *L. multiflorum*

Table 5.3. Effect of ambient sulfur dioxide on yields of representative agricultural plants[1]

	Site					
	1	2	3	4	5	6
Sulfur dioxide						
Average concentrations (ppm) during monitoring time	0.141	0.083	0.051	0.020	0.010	0 (Control)
Average concentrations (ppm) during exposure time	0.590	0.440	0.340	0.260	0.220	0 (Control)
Crop		Yield as percentage of control				
Apple	–	24.5	46.5	90.2	105.1	100
Currant	–	19.4	19.4	60.1	90.7	100
Winter rye (grain)	47.7	57.7	85.4	96.5	99.21	100
Winter wheat (grain)	42.7	55.6	85.0	92.0	98.9	100
Oats (grain)	62.4	76.1	84.7	93.0	102.4	100
Potato (tubers)	44.4	68.0	83.1	90.4	98.3	100
Alfalfa	63.8	81.0	93.1	97.4	95.6	100
Tomato (fruit)	90.2	94.4	104.8	104.6	102.0	100
Spinach	94.7	92.8	101.4	93.3	103.5	100

[1] Adapted from Guderian & Stratmann (1968).

cv S22 and *Dactylis glomerata* cv S143 were reduced when subjected to sulfur dioxide concentrations of 45–75 $\mu g/m^3$ (0.017–0.029 ppm) throughout the growing season. Although results are not always consistent, growth reductions of rye have been demonstrated by exposing plants to sulfur dioxide concentrations as little as 43 $\mu g/m^3$ (0.016 ppm) for 173 days or 106 $\mu g/m^3$ (0.04 ppm) for 194 days (Bell *et al.*, 1979). Other work suggests that timothy (*Phleum pratense*) and bluegrass (*Poa pratensis*) may be still more sensitive to sulfur dioxide (Ashenden & Williams, 1980).

Similar studies have been conducted in the USA. Although many have reported that foliar injury accompanied yield losses, some have shown significant growth and yield reductions with no visible injury. A representative exposure regime used for soya bean was 0.25 ppm (655 $\mu g/m^3$), 4 hours a day, 3 days a week for 11 weeks (Reinert & Weber, 1980).

Soya beans have been the subject of considerable study. In one open-air fumigation (Sprugel *et al.*, 1980) mean sulfur dioxide concentrations were 0.095, 0.108, 0.192, 0.255 and 0.362 ppm (about 250–950 $\mu g/m^3$ range). Exposure periods were from 6.8 to 26.1 hours. Visible injury was observed only at 0.362 ppm. Yield reductions ranged from 6.4 to 15.9% as concentrations increased.

Forests

Reductions in productivity, as well as the outright demise of the forest, are well documented around smelters and other sources of sulfur dioxide. The main concern of this discussion, however, is sulfur dioxide-related growth reduction in the absence of visible markings on the leaves. Early studies were largely concerned with severely damaged trees, such as the major study by the National Research Council of Canada (1939). Dendrochronological techniques applied in studying forests in the 1980s showed that radial growth of western larch was clearly related to annual sulfur emissions up to 15 km from the smelter (Fox *et al.*, 1986). A subsequent study was unique in attempting to relate growth responses to reconstructed sulfur dioxide exposures. While a specific threshold could not be established, a threshold effect was indicated (Kinkaid, 1987), but the extent of visible needle injury that occurred in the 1930s could not be reconstructed.

Dendrochronological studies consistently show increment growth reductions near smelters (Miller & McBride, 1975). In another approach, annual ring growth at sites near, and more distant from, a smelter in a western US Pinyon pine forest (*Pinus monophylla*) were compared. Chronologies of all sites were positively correlated with each other in years prior to operation of the smelter but not during operation (Thompson, 1981). Growth at the nearest site was considered to have been limited largely by the effluents. Two problems should be recognized: (1) no record is available as to the amount of leaf injury, if any, that was present in the past years; and (2) sulfur dioxide is not the only effluent. Heavy metals, usually in particulate form, almost always accompany the sulfur dioxide.

HEAVY METALS AND METALLOIDS

Modern control technology has eliminated much of the particulate materials from effluents of smelters and other industries. Unfortunately this does not do anything for the copper, lead, zinc, cadmium, arsenic or other substances that may have been deposited in the environs over the past century. It is important to recognize that with heavy metals and metalloids (e.g. arsenic) we are dealing with residual soil contamination as well as air pollution.

Ores contain all the chemicals found in any other body of soil, although the ores are generally higher in certain elements such as copper. When such an ore is processed to concentrate copper, zinc or other elements, everything that is not utilized, or captured by control equipment, is released into the atmosphere or water. Of these myriad wastes a few stand out as being harmful in high concentrations. The most significant of these are lead, cadmium and arsenic.

Lead

Lead occurs naturally in all soils. Typically, lead concentrations in unmineralized soils are 10–20 μg/g of dry soil. However, concentrations of 200–300 μg/g are not unusual in forested areas usually regarded as pristine (Smith, 1984). Lead in the USA and Europe continues to accumulate in forest soils far more rapidly than it flows out of the systems. Such accumulation occurs even in 'relatively' unpolluted forested environments. Near smelters, concentrations of 1000–4000 μg/g in the surface soils are not unusual. Lead concentrations adjacent to highways also can be elevated, but this problem has diminished in the USA since the introduction of lead-free gasoline.

The risk from lead is not so much to the plants, which take up very little, but to human health when affected plant parts are consumed. Lead contamination of plants is mostly from atmospheric deposition rather than absorption through the roots. Forage grasses may contain enough lead to threaten the health of grazing animals, but edible crops, being largely washed, pose little threat to humans (Ormrod, 1984).

Cadmium

Cadmium has received considerable recent attention as a medical problem causing cardiovascular and hypertension disorders (Fassett, 1975). Cadmium occurs most frequently in association with zinc ores but is otherwise among the rarest metals in the earth's crust (Lepp, 1981). Atmospheric sources are principally metal processing, fertilizers, industrial processes containing zinc, and pesticides. Sewage also is high in cadmium, and its use in soils has led to considerable study. Deposited on plants or soil, cadmium enters the food chain by uptake through the roots or leaves. Root uptake is greatest in acid soils (Bingham *et al.*, 1980).

Plant toxicity by cadmium first consists of reduced photosynthesis and transpiration rates, followed by general chlorosis (Ormrod, 1984). Such injury occurs only at concentrations well above those generally occurring even in contaminated areas, which are rarely above 5–10 μg/g of topsoil.

Normal background concentrations of cadmium in foliage are mostly well under 1 ppm on a dry-weight basis (Page *et al.*, 1981). Concentrations vary among the plant parts, being greatest in leaves. Varietal differences in uptake and sensitivity also have been demonstrated.

Arsenic

Arsenic is another potentially harmful chemical, sometimes occurring in smelter effluents and in certain coals. Since arsenic has been widely used as a pesticide and is persistent in soils, it is often found in high concentrations in agricultural soils. The natural arsenic content generally is below 40 ppm, and averages 5–6 ppm (National Academy of Sciences, 1977a; Boyle & Jonasson, 1973). But tremendous variation exists among geographical regions. Near smelters arsenic in the top few inches of soil can exceed 100 ppm (Treshow, 1970; Lepp, 1980).

Arsenic uptake by plants depends on the chemical form or solubility of the arsenic (Colbourn *et al.*, 1975). This is strongly influenced by the pH and phosphorous content of the soil. Since the different forms of arsenic vary in their availability to the plant, and hence toxicity, the form is more important than the amount in determining phytotoxicity (Ratsch, 1974). It is only the soluble arsenic that is important in causing plant injury. Consequently, it is meaningless to say that, for instance, 50 ppm arsenic will reduce growth of apple trees. Concentrations of soluble arsenic are more critical, and 1–10 ppm concentrations can injure sensitive plants.

The principal symptom caused by arsenic toxicity is the appearance of pale green to yellow lesions developing on leaves over a period of weeks. Purple to brownish spots of varying size develop with the chlorotic area, especially on older leaves. In severe cases the necrotic areas drop out, leaving a shot-hole appearance. This is quickly followed by early defoliation (Treshow, 1970). Prior to this, growth reduction can be expected along with impaired nitrogen metabolism. Elevated arsenic is associated with needle abscission and death of fine roots and sparse mycorrhizal development in certain conifers (Ormrod, 1984).

Highest plant arsenic concentrations occur in the roots. Foliar, fruit and seed concentrations vary greatly with the species, but rarely would there be over 100 ppm (dry weight), even under severely contaminated conditions. The limit set with regard to human health for arsenic content in fruits, crops and vegetables is 7 ppm (fresh weight) (US Public Health Services).

Copper and zinc

Copper and zinc in trace amounts are essential to normal plant growth. In excess they can be harmful. Both are released as particulates in stack effluents primarily from copper and zinc smelters. As particulates their dispersal is limited to the proximity of the source. With current controls in modern smelters, direct pollution from these elements is largely historical. Greater concern with copper comes from prolonged applications in fungicidal treatments. Toxicity involves interference with normal metabolic reactions, blocking specific enzymatic reactions, and ultimately suppressing growth (Ormrod, 1984; Treshow, 1970).

SUGGESTED READING

Barrett, T.W., & Benedict, H.M. (1970). Sulfur Dioxide C1-17. In *Recognition of Air Pollution Injury to Vegetation: A Pictorial Atlas* (J. Jacobson & A.C. Hill, eds). Air Poll. Contr. Assoc., Pittsburgh, Pennsylvania.

Fowler, B.A., ed. (1983). *Biological and Environmental Effects of Arsenic*. Elsevier, New York. Topics in Environmental Health, Vol. 6. x + 281 pp., illus.

Hutchinson, T.C., & Meema, K.M., eds (1987). *Lead, Mercury, Cadmium and Arsenic in the Environment*. Published for the Scientific Committee on Problems of the Environment (SCOPE) of the International Council of Scientific Unions (ICSU) by Wiley, New York. 'SCOPE 31', based on a workshop in Toronto, Canada, September 1984. xxiv + 360 pp., illus.

Lepp, N.W. (1981). *Effects of Heavy Metal Pollution on Plants*. Vol. 1. Applied Science, London. 552 pp.

Malhotra, S.S., & Blauel, R.A. (1980). *Diagnosis of Air Pollutant and Natural Stress Symptoms on Forest Vegetation in Western Canada*. Northern For. Res. Centre. Info. Rept. Nor-X-228. 84 pp.

Mislin, H., & Ravera, O., eds (1986). *Cadmium in the Environment*. Birkhauser, Boston, Massachusetts. *Experientia* Suppl. **50**; parts reprinted from *Experientia* **40**, 1984. 144 pp., illus.

Nriagu, J.O., & Davidson, C.I., eds (1986). *Toxic Metals in the Atmosphere*. Wiley–Interscience, New York. Wiley Series in Advances in Environmental Science and Technology. Vol. 17. xiv + 635 pp., illus.

Unsworth, M.H., & Ormrod, D.P., eds (1982). *Effects of Gaseous Air Pollution in Agriculture and Horticulture*. Butterworths, Boston, Massachusetts. Papers from a meeting in Sutton Bonnington, England. xiv + 532 pp., illus.

Winner, W.E., Mooney, H.A., & Goldstein, R.A., eds (1985). *Sulfur Dioxide and Vegetation: Physiology, Ecology, and Policy Issues*. Stanford University Press. A collection of 30 papers summarizing current knowledge of the effects of sulfur dioxide on the productivity of plants.

CHAPTER 6

Fluoride: Origins and Effects

Normally the lush tropical forest of the Serra do Mar above Cubatao kept the soil stable. But air pollution, most significantly fluorides from the industrial park in the Moji River Valley near Sao Paulo, Brazil, killed the vegetation on the steep slopes above the valley. On 22 and 23 January 1985, 380 mm (15 inches) of rain fell in 48 hours, causing hundreds of mudslides over an extensive area. Fluorides may have been largely controlled in the USA and Europe, but the costly control technology is not applied universally. Outdated industrial plants still release fluorides in many parts of the world.

BACKGROUND

Fluoride is ubiquitous. It is widespread in the earth's crust and is a natural component of soils, plants, animals and water. Mostly it is harmless, beneficial or even essential. But, like so many elements, it is harmful in excess. Background concentrations in the soil may reach 1%, although it is generally less than 0.05% (500 ppm) in typical agricultural soils. Soil fluoride normally is taken up by plants only in small amounts.

Polluted atmospheres provide the main source of elevated fluorides in plants. Fluoride, mostly in the gaseous HF form, is released into the atmosphere whenever the clays, rocks, coal or ores containing it are heated, or when fluoride compounds are used as a flux and allowed to escape. The high fluoride may be directly toxic to the plants or to livestock that feed on them. The ancient disease of gaddur, or gaddjax, was described in Icelandic literature 1000 years ago. Later known as dental fluorosis of cattle, the disease developed on domestic animals feeding on grass contaminated by ash deposited following volcanic eruptions (Roholm, 1937).

Direct toxicity to plants was first recorded in 1883 in Germany (in Thomas & Alther, 1966; Mayrhofer, 1893) but several reports of damage near superphosphate and glass factories followed during the next decade (Sorauer, 1914). Gases produced from open roasting at the aluminum works at Bodesberg damaged pine trees in the surrounding forests and in the forest in the Rhône Valley of Switzerland at the turn of the century (Haselhoff & Lindau, 1903). Damage to forests near superphosphate factories was also well documented (Mayrhofer, 1893; Ullmann, 1896).

With the known history of plant damage from fluorides, it is surprising that when similar symptoms were observed in US crops and forests in the 1940s the cause was accepted with such reluctance. Expansion of the aluminum industries, and of mining and processing of phosphate deposits in Florida and Tennessee, generated a tremendous increase in fluoride emissions in the USA.

Leaves of Italian prune trees, *Prunus domestica* L., developed brown spots, the margins turned brown, and they dropped early. Fruits shrivelled and many dropped two to three weeks before harvest. The location was western Washington, and symptoms appeared near every aluminum plant (Miller *et al.*, 1948). Damage was observed in 1943 and reappeared each year until effective fluoride control equipment was installed. Apricot and fig trees near an aluminum plant in California developed similar marginal leaf scorch (DeOng, 1946). Gladiolus and citrus were severely damaged in Florida.

Ponderosa pine forests near Spokane, Washington, first exhibited reddening of needles in 1943. The 'Ponderosa pine blight' area embraced some 50 square miles. Within a 3-square-mile area the forest was dead (Adams *et al.*, 1952).

Numerous, locally serious incidents appeared wherever bricks, glass or ceramics were being made or phosphate ores processed. Extensive damage and frequent lawsuits stimulated research to learn more about the effects of fluoride and the technology of how to control it.

FLUORIDE UPTAKE

Sources

The natural sources of fluoride are many. Concentrations in the earth's crust average 20–1600 ppm; 200 is typical. Igneous or volcanic rocks may contain over 4% fluoride, and phosphate outcroppings 1%. This is mostly insoluble, but some fluoride enters the soil solution and may be absorbed by plants. Absorption is greatest in acid soils, in which fluorides are most soluble. All plants contain some fluoride, typically 2–20 ppm (National Academy of Sciences, 1971). Plants in the tea family, Theaceae, may accumulate far more. Fifty to several hundred parts per million is typical (Wang *et al.*,

1949). Generally no relation exists between the fluoride content of the soil and fluoride in the plants growing in it (Fluhler *et al.*, 1981).

Rather, it is the atmospheric fluorides that are most critical to excessive accumulation and plant response. Airborne fluorides may be in gaseous or particulate forms. Particulate fluorides settle on plant foliage, where they add to the total leaf fluoride content. This is important if forage plants are involved, but rarely need concern the health of the plant since little enters the leaf.

The amount of fluoride released into the atmosphere may range from a few to several hundred pounds per day depending on the volume of production and the amount of fluoride in the raw materials, fuel or flux. Where modern control technology is utilized, emissions rarely exceed a few pounds. Several states restrict emissions to below 16 pounds (7.3 kg) per day. Once released, downwind concentrations depend on stack height, air movement and drainage, terrain and distance. The most severe fluoride damage has occurred in narrow valleys subject to inversions where air is trapped and fluoride may persist for prolonged periods. The amount of precipitation occurring that could cleanse the atmosphere, and the density of vegetation that absorbs fluoride, also influence the concentrations.

In the most extreme cases, atmospheric fluoride concentrations have exceeded 100 $\mu g/m^3$, but concentrations above 1 $\mu g/m^3$ are now unusual.* This is still enough to be harmful (Treshow, 1971).

Accumulation and concentration

The ultimate factor determining the severity of plant injury is the amount of fluoride reaching active sites and remaining there in toxic quantities.

Fluoride enters the plant primarily through the stomata and passes into the intercellular spaces where it contacts the leaf mesophyll cells. Here it is either directly absorbed into the cells or dissolved in water and transported through the vascular tissues to the leaf tips and margins where it accumulates (Compton & Remmert, 1960). Concentrations at the tips of monocotyledonous plants and needles of pine trees may be 25–100 times that at the base of the leaf. Concentrations in the margins of broad-leaved plants may be 2–10 times higher than near the midrib (Zimmerman & Hitchcock, 1956).

Once fluoride has passed through the cell wall, it reacts initially with the cytoplasmic membrane, where some fluoride may be retained. Within the cell some 60% of the fluoride is found to be associated with the chloroplasts (Chang & Thompson, 1966; Ledbetter *et al.*, 1960). Other organelles contain much of the remainder, but some fluoride enters the vacuole, where it

---

* 1 $\mu g/m^3$ of fluoride (as F^-) is equal to 0.874 ppb fluoride by weight or to 1.33 ppb by volume of any gas containing one fluoride atom per molecule. This relationship is for 35 °C and 1 atmosphere of pressure.

remains relatively inactive. It is conjectured that fluoride-tolerant species might have a capacity for accumulating fluoride in the vacuole.

Fluoride accumulates throughout the life of the leaves, so the greatest amounts occur in the oldest leaves and on older leaves of evergreen plants (Benedict *et al.*, 1964). Rapid growth will dilute the fluoride concentrations. In one study with alfalfa plants, lower leaves contained 324 ppm fluoride and the younger, upper leaves 51 ppm.

Only minute amounts of fluoride are accumulated directly in the outer bark, and virtually none is translocated from the leaves. Nor is fluoride translocated downward into the roots. Mostly, fluoride is translocated towards the older leaves. Little even gets to the flowers or fruit, which rarely have more than 2–5 ppm regardless of the exposures (Treshow, 1970).

The amount of fluoride accumulated depends in part on the characteristics of the species but mostly on the duration of exposure, concentrations in the atmosphere, and the form of the fluoride. That is, particulate fluorides may accumulate on the leaf surface but rarely enter the leaf (McCune *et al.*, 1965; Pack *et al.*, 1959). While fluorides on the leaf surface are not harmful to the plants, they can be extremely important to any livestock that may feed on them. Along this line, pastures that have been trampled, flood irrigated or otherwise exposed to soil contamination can have extremely high fluoride levels.

PLANT SENSITIVITY AND SYMPTOMS

Bio-indicators and thresholds of injury

Some plants are extremely sensitive to fluorides. These are known as bio-indicators. Injury may appear when leaf concentrations of fluoride are scarcely above background. Chinese and Royal apricot varieties, certain varieties of gladiolus, European grape, Oregon grape and Italian prune all may show injury when fluoride content of the leaves is less than 30–50 ppm. But these are exceptions.

Even the relationships between foliar and atmospheric fluoride concentrations, and injury, are general at best. No quantitative relationship exists. Too many interrelated factors influence the atmospheric concentration that is toxic, or the foliar fluoride content when injury first appears. Fluoride content merely provides a guideline or general index of possible injury and a means of measuring the effectiveness of a control program. Leaf injury, more than atmospheric concentrations or any other criteria, best represents the actual impact on the plant. Before automated monitoring equipment was developed, leaf symptoms were seriously considered as a basis for setting air quality standards (Treshow, 1965).

Despite the limitations, plants are often arbitrarily divided into sensitive, intermediate and tolerant categories. These are helpful for diagnosis, but the exposure dynamics—time of year when plants are exposed, the soil moisture regime, nutrition, moisture and temperature before, during and after exposure, age of tissues, plant vigor and, indirectly, genetic sensitivity—all influence the plant response to fluorides (Weinstein, 1977).

Many sensitivity lists prepared over the years have been based on fumigation studies and field observations. The different conditions in which the observed plants were growing explain some of the inconsistencies in the plant response. Observations are well summarized by D. C. McCune and Leonard Weinstein of the Boyce Thompson Institute (National Academy of Sciences, 1971). Responses of a few of the common species are shown in Table 6.1.

Various studies suggest that sensitive plants can be injured when leaf concentrations exceed 20 ppm (Hill, 1969; Sidhu & Staniforth, 1986). In addition to leaf necrosis such concentrations may cause reduced seed size, reductions in numbers of seeds per cone and cones per tree, and reduced germination in European larch, black spruce and balsam fir when leaf necrosis is present (Sidhu & Staniforth, 1986).

Atmospheric fluorides provide one basis for determining thresholds for plant response. D. C. McCune at the Boyce Thompson Institute assembled the available data up to 1968 that reported the gaseous concentrations and duration of exposure causing injury (McCune, 1969). The mean concentrations of atmospheric fluoride were plotted against the duration of exposure. Visible symptoms first appeared on the sensitive gladiolus varieties when exposed to concentrations above 0.5 $\mu g/m^3$ for 5–10 days. Slightly more tolerant fruit tree species and conifers first showed markings at close to 1 $\mu g/m^3$ for about 10 days. Chinese apricot, however, is at least as sensitive as gladiolus.

It can be argued that fluoride content of the tissues provides a better indicator of sensitivity since it comes closer to revealing the amount of fluoride present at active sites (Hill, 1969).

Foliar accumulation was tied to atmospheric concentration by S. S. Sidhu at the University of Newfoundland. He calculated that hydrogen fluoride concentrations of 0.3 $\mu g/m^3$ would lead to an accumulation of up to 20 ppm of fluoride in conifer foliage after 2 years exposure (Sidhu, 1980). The tremendous number of variables influencing this relationship, such as air speed, however, would make such a quantification questionable. Mitigating factors such as leaching must also be considered. Earlier work, for instance, points out the significance of the different uptake rates for each species. Data suggest that a fluoride accumulation factor might be calculated under fumigation conditions with an uncertainty factor of less than 2 (National Academy of Sciences, 1971).

Injury is rarely found on plants in the intermediate category, and even more rarely on plant species regarded as tolerant. These show symptoms

Table 6.1. Sensitivity of selected plants to fluoride

Sensitive

Apricot, Chinese and Royal
Prunus armeniaca
Boxelder
Acer negundo
Blueberry
Vaccinium sp.
Corn, sweet
Zea mays
Fir, Douglas
Pseudotsuga menziesii
Gladiolus
Gladiolus sp.

Grape, European
Vitis vinifera
Grape, Oregon
Mahonia repens
Larch, western
Larix occidentalis
Peach (fruit)
Prunus persica
Pine, Eastern white, lodgepole, scotch, Mugho
Pinus strobus
Pinus contorta
Pinus sylvestris

Pinus mugho
Pine, ponderosa
Pinus ponderosa
Plum, Bradshaw
Prunus domestica
Prune, Italian
Prunus domestica
Spruce, blue
Picea pungens
Tulip
Tulipa gesneriana

Intermediate

Apple, Delicious
Malus sylvestris
Apricot, Moorpark and Tilton
Prunus armeniaca
Arborvitae
Thuja sp.
Ash, green
Fraxinus pennsylvania var. *lanceolata*
Aspen, quaking
Populus tremuloides
Aster
Aster sp.
Barley (young plants)
Hordeum vulgare
Cherry, Bing, Royal Ann
Prunus avium
Cherry, choke
Prunus virginiana
Chickweed
Cerastium sp.
Clover, yellow
Melilotus officinalis
Citrus
Citrus sp.
Geranium
Geranium sp.
Golden Rod
Solidago sp.

Grape, Concord
Vitis labrusca
Grapefruit (fruit)
Citrus paradisi
Grass, crab
Digitaria sanguinalis
Lambs-quarter
Chenopodium album
Lilac
Syringa vulgaris
Linden, European
Tilia cordata
Maple, hedge
Acer campestre
Maple, silver
Acer saccharinum
Mulberry, red
Morus rubra
Narcissus
Narcissus sp.
Nettle-leaf goosefoot
Chenopodium sp.
Orange
Citrus sinensis
Peony
Paeonia sp.
Poplar, Lombardy and Carolina
Populus nigra L. and *Populus eugenei*

Raspberry
Rubus idaeus
Rhododendron
Rhododendron sp.
Rose
Rosa odorata
Serviceberry
Amelanchier alnifolia
Sorghum
Sorghum vulgare
Spruce, white (young needles)
Picea glauca
Sumac, smooth
Rhus glabra
Sunflower
Helianthus sp.
Violet
Viola sp.
Walnut, black
Juglans nigra
Walnut, English
Juglans regia
Yew
Taxus cuspidata

Tolerant

Ash, European mt.
  Sorbus aucuparia
Ash, Modesto
  Fraxinus velutina
Asparagus
  Asparagus sp.
Birch, cutleaf
  Betula pendula var.
  gracilis
Bridalwreath
  Spiraea prunifolia
Burdock
  Arctium sp.
Cherry, flowering
  Prunus serrata
Cotton
  Gossypium hirsutum
Current
  Ribes sp.

Elderberry
  Sambucus sp.
Elm, American
  Ulmus americana
Juniper (most species)
  Juniperus sp.
Linden, American
  Tilia americana
Pear
  Pyrus communis
Pigweed
  Amaranthus retroflexus
Plane tree
  Platanus sp.
Plum, flowering
  Prunus cerasifera
Pyracantha
  Pyracantha sp.

Squash, summer
  Curcurbita pepo
Strawberry
  Fragaria sp.
Tomato
  Lycopersicon esculentum
Tree of Heaven
  Ailanthus altissima
Virginia creeper
  Parthenocissus quinquefolia
Willow (several species)
  Salix sp.
Wheat
  Triticum sp.

---

only when foliar fluoride content would be several hundred parts per million and atmospheric concentrations well above 1 ppb persist.

Based on the appearance of foliar lesions, some interesting although very broad generalizations can be drawn. Perhaps most significantly, conifers important in forestry are among the more sensitive species; broad-leaved species tend to be more tolerant. However, certain varieties of fruit trees are highly sensitive. Many other fruit and berry crops are at least moderately sensitive. Vegetable and field crops generally are among the most tolerant. Ornamental plants vary tremendously in sensitivity, with gladiolus and tulips among the most sensitive.

The fluoride concentrations found in injured leaves vary greatly, especially in the more tolerant plants, in which concentrations may exceed 1000 ppm. Foliar fluoride concentrations exceeding 100 ppm are common in areas where fluoride sources lack controls. Typical concentrations often range between 100 and 500 ppm.

When leaf injury is severe, growth is suppressed and the tree may ultimately die. But in the absence of necrosis or chlorosis, growth is generally unaffected. If anything, growth may be stimulated by fluoride concentrations slightly above background (National Academy of Sciences, 1971; Treshow, 1970). Somewhat higher concentrations may cause growth reductions before any visible markings appear. The concentration at which this might be expected would be in the 1–5 $\mu g/m^3$ range depending on the species sensitivity and specific growing conditions.

The atmospheric concentrations and rate of exposure, i.e. the dose, strongly influence the appearance of symptoms. Given optimal conditions, sensitive plants may be marked by exposure to hydrogen fluoride concentrations below 1.0 ppb (McCune, 1969; Treshow & Pack, 1970).

Plant symptoms and response

Marginal leaf necrosis! This classic symptom of fluoride toxicity is rare now in North America and Europe. But leaf necrosis still reveals the presence of elevated fluorides where it exists. The earliest sign of injury is the appearance of water-soaked, dull, dark-green tissue along the leaf tip and margin. The symptom may appear within 24 hours of exposure. Generally though, fluoride concentrations build up over a period of days or weeks, and symptoms appear only when some critical threshold has been reached, often coupled with stress from high temperature or drought. The water-soaked stage may be brief or lacking. Most often in the field, the first sign of injury is the appearance of light- to dark-brown necrotic lesions near or at the leaf tip. Lesions may range from a few millimeters to over a centimeter across. Such symptoms are especially characteristic on sensitive plants such as apricot, prune and grape (Figure 6.1). When fluoride concentrations are extremely high, fluoride apparently accumulates at random locations in the leaf faster than it can be translocated, and scattered intercostal lesions develop. Chlorosis, or leaf yellowing, is often associated with the higher concentrations. These are sometimes referred to as chronic symptoms.

A distinctive feature of the necrosis is the presence of a sharply defined reddish-brown band a millimeter or so wide separating the necrotic from healthy tissue. The band often consists of an abscission layer formed by cell division, so that the necrotic area drops off readily. With successive exposures to fluoride, new necrotic bands develop. If a necrotic band does not develop quickly, bands of reddish tissue separate the necrotic zone, giving the injury a wavy, zonate appearance. At the extreme, over 90% of the affected leaves may become necrotic, and over 90% of the leaves of a tree may show injury. Most typically, however, injury appears on less than 5% of the leaves, and less than 10% of an affected leaf is necrotic. Even under extreme conditions, the injured leaves tend to remain on the tree, although defoliation has been reported as a symptom (Brewer, 1960). The most sensitive leaves on stone fruit trees are those on sucker shoots with their softer, rank growth.

Reddish-purple discoloration, developing along the margin and progressing intercostally in a mottled, purplish pattern, is a distinctive symptom on lilac. This anthocyanin excess, or masking of a loss of chlorophyll, also may develop on Oregon grape and a few other species. Ultimately, affected tissues become necrotic.

Toxicity on conifers is less definitive. The reddish-brown needle tip necrosis appearing on pines is a stress response characteristic of a number

Figure 6.1 Marginal leaf necrosis on apricot, representing the classic symptoms of fluoride injury

of pathogens, including winter injury and drought as well as other air pollutants. The tip necrosis on fir and spruce is more of a dull brown in color, but is also too general a response to be totally diagnostic. In such cases fluoride analysis of the needles, presence of a fluoride source and other background information is helpful. It is especially important to survey other sensitive species for symptoms (Treshow, 1970).

Fluoride injury on conifers begins at the needle tips and progresses basally, typically leaving a succession of reddish-brown bands across the paler brown necrosis (Figure 6.2). Needles are most sensitive as they emerge from the fascicle in the spring and elongate over the next few weeks. They become progressively more tolerant with the season. Needles formed in previous years are more tolerant and rarely injured by current exposures.

Chlorosis is a characteristic response on some species, notably citrus. The chlorosis begins along the margin and extends between the larger veins. The pattern is reminiscent of certain nutrient deficiencies. The early pale green gradually intensifies and may turn completely yellow or, finally, necrotic. The demarcation between healthy and chlorotic tissue becomes more sharply defined.

Symptoms on gladiolus and other monocotyledonous plants are much the same as on the broad-leaved plants. Necrosis begins at the leaf tips, progresses downward, and reddish-brown bands separate and denote zones of exposures. While this development is characteristic on gladiolus, false Soloman's seal, tulips, iris and similar plants, it is not characteristic of corn. Fluoride injury on corn consists of a chlorotic mottle or flecking concentrated along the leaf margin and towards the leaf tip (Figure 6.3). When injury is most severe the entire leaf may be extremely chlorotic, sometimes interspersed with small islands of green tissue. Veinal areas also tend to remain green. Similar symptoms appear on sorghum, but there is a tendency

Figure 6.2 Necrosis of pine needles with darker brown banding characteristic of fluoride injury

Figure 6.3 Chlorotic stipple characterizing fluoride injury on moderately sensitive corn

for dark, reddish-brown spots to develop near the leaf tip and margins (Treshow & Pack, 1970).

Fruits provide a fascinating response to fluoride. Mostly they are unaffected, but there are classic, highly noteworthy exceptions. Soft suture or suture red spot of peach is best known (Figure 6.4). The disease was first described and given the name fluoride suture in the 1930s, when fluoride compounds were tested for insect control (Anderson, 1931). Symptoms are

Figure 6.4 Soft suture (suture red spot) of peach attributed to fluoride, showing the prominent reddening and softening of tissues at the basal end of the fruit

characterized by premature reddening and ripening of a local, rather sharply delimited area along one or both sides of the suture line towards the lower third of the fruit. The premature ripening may involve a local swelling and softening of the affected area. This portion may then be overripe and sunken when the rest of the fruit ripens. Most distinctively, in the early stages, the flesh beneath the skin turns red while surrounding tissue remains normal. Symptoms may appear two to three weeks before harvest. The condition is most severe when the crop is light.

Suture red spot is an indirect response to fluoride in that it is related to boron and calcium nutrition. Even in cases where leaves are not marked by fluoride, the disorder has rendered as much as 90% of the peach fruits unmarketable. During the pit-hardening stage of fruit maturation the need for calcium is greatest. Fluoride disrupts the normal calcium and boron metabolism, a condition readily corrected by a single calcium chloride spray applied when the pit begins to harden (Benson, 1959). The interactions are more complicated than this since spray applications of boron have also been shown to reduce suture red spot (Weinstein & Alsher-Herman, 1982).

Basically similar symptoms occur on other fruits. A local necrosis has been described at the stylar end of apricot, cherry and pear fruits in France (Bolay and Bovay, 1965). It has also appeared on cherry fruits, on which it was called shrivel tip (Treshow, 1970).

Fluoride may adversely affect production once visible foliar symptoms appear. There is some evidence that growth of conifers may be impaired before any visible symptoms appear (Treshow *et al.*, 1967; Keller, 1977). Airborne fluorides have markedly reduced yield of Valencia oranges when chlorosis was prominent, foliar fluoride concentrations approached 200 ppm, and mean atmospheric concentration exceeded 6 ppb (Leonard & Graves, 1966). In other studies concentrations in the 6 ppb range also caused significant reductions in weight of tomato fruits. Foliar content was 1252 ppm and considerable leaf necrosis appeared (Pack, 1966).

Pollen tube growth and receptivity may be impaired on some plants (Sulzbach & Pack, 1972). Other studies involved cherries (Facteau & Rowe, 1977). There is at least one publication pointing to a stimulating effect of fluorides on pollen tube growth (Lai Dinh *et al.*, 1973). Work at the University of Utah using culture dishes showed stimulation at 10^{-4} M sodium fluoride (Treshow, unpublished). The potential impact of such effects on actual production has yet to be established, although some inhibition of fruit set may have resulted at atmospheric concentrations above 2 ppb (Facteau *et al.*, 1973).

Diagnosis and mimicking symptoms

Diagnosis of fluoride injury is often complicated by the presence of essentially similar symptoms caused by agents that may be totally unrelated to fluoride. The principal means to identify the fluoride-related symptoms is by observing fluoride-sensitive species in the area and knowing the relative sensitivity of the plants in question. If symptoms appear on tolerant species, fluoride is not likely to be involved. Fluoride analysis of the foliage also is helpful, although if concentrations are in the range of 20–50 ppm diagnosis can still be in doubt. Also, even very high concentrations do not assure that fluoride is responsible for the injury. Similarly, the presence of a potential fluoride source is not always relevant.

Still, the amount of fluoride in leaves can be helpful to diagnosis. Since background concentrations are quite low, accumulations of over about 20 ppm in foliage generally indicate the presence of atmospheric fluoride. Of course, whether or not damage results is another matter; it depends on the sensitivity of the plant and the environment. Familiarity with the details of symptomatology is extremely important for diagnosis. It requires considerable experience not only with fluorides but also with plant responses to other stresses and disorders.

Marginal necrosis and tip burn are ubiquitous responses to almost any stress. There are nuances that may be suggestive of a pollutant but these are not always definitive. Moisture stress and low temperatures are most often responsible for mimicking symptoms. This is especially true with pine trees and other conifers. Questions often can be resolved by obtaining adequate background information. Other pollutants that might mimic fluoride injury include sulfur dioxide and chlorine. Pesticide toxicity also may cause problems in diagnosis. Where chlorosis occurs on broad-leaved plants, nutrient deficiency symptoms, mostly from zinc or manganese, can be similar to fluoride toxicity. In all cases knowledge of the total syndrome facilitates an accurate diagnosis.

MECHANISMS OF ACTION

The leaf necrosis and chlorosis caused by fluoride are obvious, but what precedes the appearance of these symptoms? What are the mechanisms of toxicity? How significant are subliminal effects prior to the appearance of visible injury?

Fluorides, as with other gases, first pass through the stomata. The guard cells are the first to be affected (Poovaiah & Wiebe, 1973). However, the partial closure induced is not the most sensitive response. Rather, it appears to be the chloroplast, and most specifically the thylakoid, that is most sensitive (Zwiazek, 1987; Miller & Miller, 1974). The thylakoids become swollen and curled, finally becoming undulating structures. An increase in the size and number of plastoglobuli has also been observed (Soikkeli & Karenlampi, 1984). Fluoride also causes the appearance of lipid materials in the cytoplasm during early stages of injury that suggests damage to the membrane. This is correlated with decreases in phospholipids and increases in neutral lipid levels in fluoride-treated seedlings (Zwiazek, 1987).

It is not surprising that metabolic effects are associated with these ultrastructural changes. Photosynthesis has been especially well studied and has long been recognized to be inhibited by fluoride, even though the effect appears to be transient (Thomas & Hendricks, 1956). Concentrations of fluoride in the 1–12 ppb range can cause depressions in the amounts of chlorophyll a and b; again the effect is transient unless injury has progressed to the stage of visible chlorosis (Weinstein, 1961).

Measurements of carbon dioxide exchange have been widely used to study air pollutant effects. The inhibitions of carbon dioxide uptake caused by fluoride may be due to the inhibition of enzymes or coenzymes with metal components involved in utilizing light energy (Spikes et al., 1955). The addition of magnesium was found to neutralize the inhibitory effect of fluoride, which suggests that fluoride may be rendering the magnesium, essential to thylakoid function, inactive (Ballantyne, 1972).

The mitochondria and respiration are clearly less sensitive to fluoride than is the chloroplast. Depending on the concentration, respiration may be either stimulated or inhibited (Miller & Miller, 1974). Many variables influence this, including the age of the tissues, the nutrient status and the exposure regime. Mitochondrial phosphorylation was severely inhibited by fluorides. This apparently was related to the sensitivity to fluoride of a key phosphorylating enzyme, adenosine triphosphatase. The response was associated with mitochondrial swelling and leakage of proteins that suggested a direct effect on the membranes. Inhibition of mitochondrial malic dehydrogenase only in vivo also suggests that the effect of fluoride may be more on the membrane than on the enzyme (Psenak *et al.*, 1977). The relative proportion of the pentose phosphate pathway over glycolysis in leaves is increased by fluoride. This may be caused by the inhibition of enolase and glycolysis, which leads to an increased flow of carbon through the pentose pathway (Weinstein & Alscher-Herman, 1982).

Fluorides are best known for their effect as enzyme inhibitors, so a number of metabolic processes might be affected (National Academy of Sciences, 1971). Reported changes in sugars, polysaccharides and organic acid concentrations could all be associated with such inhibition. Enzymes such as glucose-6-phosphate dehydrogenase, enolase and phosphoglucomutase have been been shown to be affected by fluoride (in Malhotra & Khan, 1984). Inhibition of phosphoglucomutase may account for increases in concentrations of non-reducing sugars. This in turn could account for the decrease in starch concentrations (Koziol, 1984).

Changes in keto acids, organic acids, amino acids and amides, free sugars, peroxidase, DNA and RNA phosphorus, and starch and non-starch polysaccharides have all been found in plants exposed to fluorides (National Academy of Sciences, 1971; Treshow, 1984).

It may be that one of the underlying mechanisms of fluoride action may be in interfering with metal-ion transport and/or inactivating metal ions at their sites of physiological activity. This could be inferred from the striking resemblance of chronic fluoride symptoms to deficiencies of magnesium, manganese and zinc nutrients (Weinstein & Alscher-Herman, 1982).

As with many other pollutants, it appears reasonable to conclude that fluoride acts first on the thylakoid membranes, presumably the protein component. This may bring about a number of secondary reactions. But it is equally likely that a number of enzymes are affected directly, thus inhibiting several reactions. Interactions with bivalent ions, e.g. magnesium, may be especially critical. Any or all of these biochemical effects go on to cause the visible changes and symptoms known to result from fluoride toxicity.

SUGGESTED READING

Murray, F., ed. (1982). *Fluoride Emissions: Their Monitoring and Effects on Vegetation and Ecosystems*. Academic Press, New York. Papers from a workshop in Sydney, Australia, August 1981. xiv + 234 pp., illus.

National Academy of Sciences (1971). *Fluorides*. Biological Effects of Atmospheric Pollutants. NAS, Washington DC, 295 pp.

Weinstein, L.H. (1977). Fluoride and plant life. *J. Occup. Med.*, **19**, 49–78.

CHAPTER 7

Smog: The Discovery of PAN as a Photochemical Pollutant

In the summer of 1945, farmers in the Los Angeles area were asking plant pathologists at the University of California a frequent question: 'What's wrong with my plants?' Usually they had an answer. This time they didn't. The pathologists had never before observed the silvery, glazed leaves of lettuce and spinach plants that rendered crops unmarketable (Figure 7.1). Neither had any of the other experts who were called in. Nor did a search of the literature disclose any answers.

In an effort to determine the cause of this new disease, at first named 'silver leaf', the investigators asked a number of questions. These were questions basic to any diagnosis. Were there any organisms consistently associated with the symptoms? Insects? Fungi? Bacteria? Abiotic stress from nutrient disorders or climate? What crops were affected? Were the symptoms associated with any particular growing condition? Fertilizing practices? Irrigation? Were affected plants limited to certain soils? What was the cropping history? When did the symptoms first appear? And finally, what was the geographical distribution of the affected plants, and had there been any unusual weather conditions?

This last question provided a major clue. The glazed plants were found only in parts of the Los Angeles Basin and only following periods of 'smog'. Smog was well known in Los Angeles when the symptoms first appeared. The term, which was first used in London in 1905 to denote a combination of smoke and fog, was applied in the 1940s to the acidic, eye-irritating haze that now concerned residents of Los Angeles. But was it damaging to plants? Perhaps some toxic chemical was present in the smog. But what? Sulfur

Figure 7.1 Bronzing of Great Lakes lettuce following characteristic injury appearing in Los Angeles lettuce fields in the 1940s

dioxide was suspected, but the symptoms found in Los Angeles were not at all characteristic of injury caused by this chemical. Could some other air pollutant be involved? At the time, sulfur dioxide was the only widely recognized pollutant.

The Los Angeles atmosphere was tested during days when visibility was poor and the results compared against analyses made during days when visibility was good, i.e. over 7 miles (Air Pollution Control District, 1951). As expected, differences in concentrations of chemicals were considerable. Sulfur dioxide and sulfuric acid concentrations were strikingly greater when visibility was poor, and these pollutants were not discounted. With over 500 tons of sulfur dioxide released over the area each day, and hourly concentrations exceeding 0.1 ppm (260 $\mu g/m^3$), it would be surprising if some effects did not occur.

Carbon monoxide concentrations also were elevated, but this chemical is not harmful to plants. Oxidants, aldehydes, hydrocarbons, nitrogen oxides and ozone, also elevated during smoggy days, were more suspect. But were any of these harmful to plants, and if so, did they cause the characteristic glazing of leaves?

The next step was to identify the specific toxic chemicals and then to expose plants to them. Given the technology of the day, identification was not easy. As an example of the 'state of the art', ozone was measured by observing the depth and extent of cracking of small strips of rubber (Bradley and Haagen-Smit, 1950).

In 1949 fumigation studies were begun at the California Institute of Technology Earhart Plant Research Laboratory in which plants were exposed to a number of chemicals. But none produced the characteristic symptoms. Since the polluted air had a highly oxidizing characteristic and contained large quantities of unburned hydrocarbons, Dr A. J. Haagen-Smit (1952) of the California Institute of Technology postulated that toxic materials must be products of oxidation of hydrocarbons. Chemists learned that ether-soluble chemicals extracted from filters exposed to smog could be simulated artificially by adding ozone to gasoline exhaust vapors. It was only when gasoline vapors were combined with ozone that the fumigations produced promising, similar, symptoms. Introducing ozone into the exhaust stream of a small engine became the standard means of exposing plants in chambers to synthetic 'smog'. By exposing plants to smog under such relatively controlled conditions, researchers confirmed that some chemical associated with auto exhausts caused the symptoms. As in the field, spinach, endive, romaine (cos) lettuce and mustard were among the most sensitive plants.

Which of the chemicals in the exhaust were the toxic components? Ed Stephens and his colleagues at the University of California Citrus Experiment Station at Riverside (Stephens *et al.*, 1956) analyzed smog reaction products generated with mixtures of nitrogen dioxide and organic compounds in a new 250–500 m infrared cell. They observed that ozone was formed as nitrogen dioxide disappeared. They found another reaction product from pentenes that was considered particularly important, but it could not be identified. They called it Compound X. A few years later it was found to consist of a group of peroxyacyl radicals which in turn could add nitric oxides, forming peroxyacyl nitrates or PANs.

Peroxyacyl nitrates are a homologous series of compounds. The first member, peroxyacetyl nitrate (PAN), appears to be the principal constituent causing plant damage. A second member, peroxyproponyl nitrate (PPN), is about four times more toxic, but occurs in much lower concentrations not harmful to plants. Peroxybutyl nitrate (PBN) is about eight times as toxic as PAN, but has not been found to occur in ambient air. All produce the same symptoms (Darley *et al.*, 1966). These were the chemicals causing the oily glazing on leafy vegetable crops. Within this family of compounds PAN is the most abundant.

SYMPTOMS AND SENSITIVITY

Once PAN had been synthesized, the injury symptoms it caused under controlled conditions proved to be indistinguishable from those found on the leafy vegetable crops in the fields of California. By 1955 such symptoms also had been reported in at least 18 other states as well as near larger cities of many countries around the world (Went, 1955).

Peroxyacyl nitrates typically injure the lower leaf surface of affected plants. Leaves of such sensitive plants as romaine (cos) lettuce, Swiss chard, spinach and celery develop a silvery or bronzed glazing on their lower surfaces (Table 7.1). The upper leaf surfaces may be similarly marked following severe exposures to higher concentrations. These symptoms are most clearly illustrated in *Recognition of Air Pollutant Injury to Vegetation: A Pictorial Atlas* (Taylor and MacLean, 1970).

The overall appearance of PAN injury in the field varies somewhat with each crop. Some type of glazing or bronzing distributed irregularly over the under-surface of the leaf best characterizes smog injury on herbaceous plants (Middleton *et al.*, 1950, 1953). Symptoms may be limited to a water-soaked appearance, as in dahlia; silvering or bronzing of the lower leaf surface, as in spinach, bean or petunia (Figure 7.2); browning or bronzing of either

Table 7.1. Sensitivity of plant species to PAN-type injury[1]

Sensitive		Resistant	
Crops			
Spinach	Beet	Cabbage	Onion
Endive	Corn	Cauliflower	Corn
Oats	Celery	Rhubarb	Cucumber
Romaine lettuce	Pepper	Carrot	Strawberry
Swiss chard	Tobacco	Squash	
Alfalfa	Clover		
Beans			
Ornamental plants			
Petunia	Fuchsia	*Anthurium*	*Coleus*
Mimulus	*Impatiens*	Bromeliad	Cyclamen
Snapdragon	Mint	*Calendula*	Ivy
Primrose	*Ranunculus*	Camellia	Narcissus
Aster		Carnation	Lily
			Most house plants
Weeds			
Annual bluegrass	Mustard		
Pigweed	Jimson weed		
Chickweed	Dock		
Wild oat	Ground cherry		

[1] Adapted from Noble (1965).

Figure 7.2 Characteristic glazing of the lower surface of petunia leaves produced by Los Angeles smog in the 1950s. (Photo courtesy of Wilfred Noble.)

surface, as in romaine (cos) lettuce, zinnia and some chrysanthemum varieties; chlorotic or brownish mottling, as in tomato, pumpkin and mallow; reddening caused by an increase in anthocyanin, and cork formation as in beets; tan banding as in barley and annual bluegrass; or longitudinal streaking as in oats, corn and some grasses.

Bronzing and glazing appears on both surfaces of endive, a species with palisade parenchyma on both leaf surfaces. However, interveinal silvering and bronzing is most frequent and prominent on the lower surface.

Silvering can vary from a glistening sheen on spinach to an almost oranging on table beet or milky whiteness on snapdragon. Small plants and young leaves are most sensitive. Bronzing ranges from a glossy, bronze sheen to general necrosis.

The precise nature of the symptoms varies depending on such environmental factors as tissue maturity, concentrations, duration of exposure and the light regime. Because of the variations in symptoms, it is preferable to refer more broadly to 'PAN-type' oxidant injury (Jaffe, 1966). Conditions conducive to the production of a succulent, fast-growing plant increase the sensitivity of the plant to PAN damage. Thus, it is the young, actively expanding tissues of sensitive plants that become visibly marked. Since the specific region of the most rapidly expanding tissue of a leaf is the most sensitive, maturing bands of tissue across the leaf tend to be injured by a given exposure. Repeated exposures thus tend to produce bands of glazed or necrotic tissue (Figure 7.3). If exposures are prolonged, the bands coalesce and a broader affected area appears so that any banding is obscured, and the typical, general glazing appears. Banding is most prominent on leaves of petunia, ryegrass and annual bluegrass, but also appears frequently on such sensitive broad-leaved plants as spinach, beet, chard, *Mimulus*, chickweed, pigweed and dock. While banding appears on broad-leaved species when the period of exposure is of fairly short duration, this expression is most obvious on susceptible, longer-leaved grass species.

Apparently the cells that have just finished differentiation are most severely injured (Glater *et al.*, 1962). These probably are most sensitive because of their greater metabolic activity, but also because of the relatively greater tolerance of other cells. Stomata in the young portion of the leaves are not fully formed; stomatal initials are still differentiating, and intercellular spaces are undeveloped. The resistance of older tissues is attributed to the

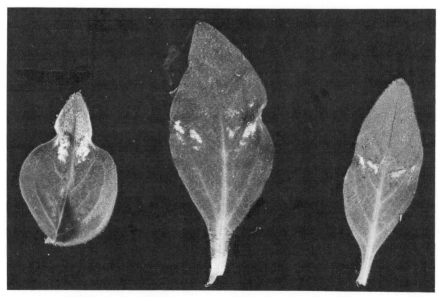

Figure 7.3 PAN injury to petunia showing simultaneous injury to tip of the youngest leaves and center of older leaves. (Photo courtesy of B. L. Richards.)

normal development of a waxy layer around individual cells as well as greater cuticle thickness, lignification and reduced stomatal activity. Reduced gas exchange rate and metabolic activity, and possibly the reduced gas exchange rate and metabolic activity of older tissues, may all increase tolerance.

The same mechanism explains the transverse banding expression found in monocotyledonous plants. These plants possess a persistent basal meristem. Since new cells are formed at the leaf base, cells mature from the leaf tip down, and sensitivity of cells varies with the distance from the meristem. The young basal cells and oldest tip cells are most tolerant. Cells a short distance from the meristem are most sensitive. Following fumigation a single discrete band of cells will be injured. Following successive fumigations a broader band, or even longitudinal streaks, will develop as the leaf matures.

This banding expression is one of the best means of identifying PAN injury and separating it as a causal agent from ozone, which may on occasion produce similar glazing symptoms. On species where PAN symptoms are not specific, it is only possible to state that the injury is the result of photochemical air pollution.

PATHOLOGICAL HISTOLOGY

Photochemical pollutants, including PAN, enter the leaf through the mature, functional stomata. Once within the substomatal chambers, the pollutant attacks the mesophyll cells bordering the intercellular spaces. These morphological and histological effects of ozonated olefins, and later of PAN, were first studied by Ruth Bobrov at UCLA (Bobrov, 1952; Glater *et al.*, 1962). She found that the earliest visible indication of injury was the oily, shiny, water-soaked appearance of sensitive tissues on the lower leaf surface. By the time this appears, tissue alterations already have begun. Tiny, raised blisters appear that are formed by the swelling of guard cells and other cells nearest the stomata. These become engorged with water and increase in width, causing the stomata to enlarge further. If the guard cells become excessively distorted, they will be permanently injured and collapse, closing the stomata and preventing further entrance of PAN.

By the time the epidermal cells collapse, the entire leaf becomes turgid. Permeability may be disturbed, so that excessive water enters the affected cells. Stretching of cells as they engorge with water gives the underside of the leaf its shiny, water-soaked appearance. If the fumigation does not persist, or is not too severe, the turgid cells may recover after a few hours, leaving no trace of visible injury.

Cells of the spongy mesophyll nearest the intercellular spaces, or substomatal cavity, are affected first. Small, electron-dense granules appear in the chloroplast stroma soon after fumigation (Thomson *et al.*, 1965).

Subsequently 'crystalline' arrays of granules appear, and the shape of the chloroplast is altered. The granules seem to fuse into rods and then into an organized system of plates that persist and even seem to continue to develop. Finally the integrity of the chloroplast is lost and the membranes are disrupted. As the chloroplasts break down and become dispersed into the cytoplasm, the entire protoplast aggregates into a large mass that condenses in the interior of the cell, moving away from the cell walls and causing the cell to collapse. The plasmodesmatal connections with neighboring cells appear to persist. While the lower surface is usually injured first owing to the larger amount of intercellular space, the reverse also may be true. The tissue that is injured appears to depend to a great extent on its age. Palisade cells differentiate later than spongy parenchyma and may sometimes be in less sensitive condition at the time of exposure. Smaller veins may be damaged together with the mesophyll, but the larger veins and midrib usually remain intact.

Monocotyledonous plants, including oats, corn, barley and grasses, appear dark green as if water were trapped beneath the epidermis. As cell damage progresses, the dark green water-soaked areas develop into yellow streaks that follow the zones where stomata are most dense. The yellow soon turns to brown, and longitudinal necrotic streaks appear between the larger veins.

In all of these plants damage seems to be partly proportional to the amount of internal air space existing within the leaf. Other factors that influence sensitivity include the plants' inherent susceptibility or immunity; their metabolic activity; the age of the plant, the leaf and leaf tissue; and many environmental factors such as temperature, soil moisture, light and nutrition.

BIOCHEMICAL AND PHYSIOLOGICAL EFFECTS

Once the cause of symptoms has been determined, it is desirable to understand more fundamentally what caused them and how symptoms arose. Also, what adverse effects might have occurred before visible injury appeared? What are the fundamental biochemical and metabolic processes that led to the symptoms?

As with many other pollutants, PAN can interfere with a number of basic metabolic reactions. The problem is, which interference comes first, and which is most critical? We might first look at how PAN is incorporated into the plant constituents so we can learn which metabolic pathways might most likely be affected. The research group at the University of California synthesized PAN labeled with carbon-14 so that its pathway in the plant might be followed (Stephens *et al.*, 1961). When the cells were fractionated, much of the carbon-14 appeared in the chloroplasts, so it is here one might look first for harmful effects. But in what part?

Of all plant reactions, photosynthesis taking place in the chloroplasts must be regarded as one of the most critical. It also proves to be one of the most sensitive to PAN. Exposure to even low concentrations of PAN for brief periods was found to inhibit evolution of oxygen in chloroplasts, although no effect on photophosphorylation was evident (Duggar *et al.*, 1965). When isolated chloroplasts were treated with PAN, electron transport, carbon dioxide fixation and photophosphorylation were all inhibited (Coulson and Heath, 1975).

The inhibitory effect may be attributable to the ability of PAN to oxidize critical sulfhydryl (SH) groups in proteins and metabolites such as cytosteine, reduced glutathione, coenzyme A (CoA), lipoic acid and methionine (Mudd, 1975). NADH and NADPH, so vital in oxygen transport, also can be oxidized by PAN, an effect most likely attributable to diverse effects on the enzymes concerned. The basic mechanism, and one that helps explain the observation that plant damage from PAN occurs only in light, seems to be the sulfhydryl oxidation. Enzymes containing free sulfhydryl groups for catalytic action are especially sensitive to PAN. The enzymes concerned are located in the stroma of the chloroplast, and it is here that the first signs of damage are detectable with the electron microscope as well as in histological, microscopic studies (Mudd, 1963; Thomson *et al.*, 1965).

Although photosynthesis appears most broadly to be adversely affected by PAN and the protein thiols more specifically, PAN also reacts with low molecular weight thiols. CoA was one of the thiol compounds tested (Mudd & McManus, 1969). When products were separated by ion-exchange chromotography, CoA disulfide was the major product. This was also the case when CoA was oxidized with hydrogen peroxide. Finally, it should be noted that damage may be prevented by adding sulfhydryl reagents, that protect the sulfhydryl group, to enzyme reactions mixtures, chloroplast and mitochondrial suspensions, or spraying them on intact plants.

CONCENTRATIONS CAUSING INJURY

It is difficult to say, in the field, exactly what concentrations of any chemical are harmful to plants. This is in part because many other chemicals also may be present that have additive, or even synergistic, effects. In the case of PAN, dose responses also are complicated by the limited data available on ambient concentrations. The available data were reviewed by O. C. Taylor in 1968, who noted that concentrations as high as 210 ppb had been reported in Los Angeles and 58 ppb in Riverside, California. Concentrations up to 54 ppb were recorded in Salt Lake City, Utah, in 1968. Concentrations regularly range from 10 to 20 ppb in these, and presumably other urban, areas (Darley *et al.*, 1968), but little or no recent data are available.

Acute injury on such sensitive plants as petunia and tomato can be caused by 4-hour exposures to the PAN concentrations found in such ambient air (Taylor, 1969). Fumigation with synthesized and purified PAN required higher concentrations to be injurious. Frank Anderson, at the University of Utah (Treshow, 1970) found that PAN concentrations from 20 to 100 ppb for periods of 1–2 hours caused visible injury to alfalfa and other sensitive species. Concentrations above 100 ppb caused complete collapse of tissues. Below 100 ppb, visible symptoms consisted mostly of chlorosis.

GROWTH AND REPRODUCTION

We know less about the effects of PAN on growth and reproduction than any other major pollutant. The reasons are clear. PAN is relatively difficult to generate and monitor, and it is dangerous to work with. Despite precautions, its explosive nature has led to serious accidents almost wherever PAN has been studied—California, Pennsylvania, New Jersey and Utah. Also, about the time techniques for production studies were becoming sufficiently sophisticated, the overwhelming importance of ozone became recognized, and researchers shifted their emphasis to this more ubiquitous and more readily monitored pollutant.

Studies of growth responses to PAN were conducted at a time when the main approach was to make comparison of plants grown in chambers subjected to ambient versus filtered air. This is appropriate when there is only a single pollutant to filter out. Unfortunately this was not the case in Los Angeles, where PAN occurred in combination with ozone and a number of other pollutants.

Consequently, varied studies conducted, demonstrating that avocado plants, Kentia palm and, most significantly citrus, grew better in filtered than ambient air, do not tell us the precise role of PAN as opposed to the impact of ozone or other atmospheric pollutants present (Thompson, 1968; Todd *et al.*, 1956).

SUGGESTED READING

Robert Guderian, ed. (1985). *Air Pollution by Photochemical Oxidants: Formation, Transport, Control, and Effects on Plants*, Springer-Verlag, New York. Ecological Studies, Vol. 52. xii + 346 pp., illus.

CHAPTER 8

Ozone as an Air Pollutant

OZONE FORMATION AND TRANSPORT

Air pollution is not a local problem. This is especially true of ozone. This molecule, made up of three oxygen atoms, is often thought of as beneficial. In the upper atmosphere, where it is a natural compound that helps to screen us from harmful effects of ultraviolet radiation, it is. Ozone of concern here, though, is that produced approximately at ground level by complex photochemical reactions involving sunlight and emissions of hydrocarbons and nitrogen oxides from motor vehicles and other sources of high-temperature combustion.

Although other photochemical oxidants such as PAN can play a role in causing vegetation injury, and can impair human health, it is the ozone that is now recognized to be primarily responsible for the adverse effects.

The precursors of ozone are widespread and their movement even broader. Ozone in the troposphere is formed when nitrogen dioxide (NO_2) is converted to nitric oxide (NO) by the action of sunlight. The freed oxygen atom reacts with oxygen molecules (O_2) to form ozone (O_3). In the absence of competing or scavenging molecules, the reaction reverses to produce a state of equilibrium between the ozone, nitrogen dioxide and nitric oxide. But when organic molecules, largely the volatile organic hydrocarbons (VOCs) are present, they react with the nitric oxide, stopping the back reaction so that ozone accumulates. Other molecules, notably the hydroxyl radical (OH) and its precursors, are also important. Aldehydes and other automobile exhaust products are important sources of hydroxyl radicals. Natural hydrocarbons emitted by vegetation similarly undergo photo-oxidation reactions to enhance ozone formation, although they can also serve as scavengers of ozone. An excellent, complete, yet simple, review of the related atmospheric chemistry is given by Alan Wellburn (1988).

Several chemicals are involved in photochemical pollution. Among these is PAN (peroxyacetyl nitrate), which persists in the atmosphere and contributes to the long-range transport of the oxides of nitrogen (NO_x). Hydrogen peroxide (H_2O_2) also plays a role in photochemistry. Its major importance arises from its high solubility in water, being taken up by aqueous droplets and potentially playing a role in acidification of rain. Alone, it can injure plants.

Atmospheric stability, wind speed, sunlight, temperature, general weather conditions, concentrations of the precursors, mixing, scavenging during transport and injection of new emissions all have a strong influence on the formation and transport of ozone and the downwind ozone concentrations. Incursions of ozone from the stratosphere may also play a role; this source is generally considered to be minor compared with tropospheric ozone formation, but some consider photochemically generated stratospheric ozone to be significant (Altshuller, 1986). Stratospheric ozone is generally thought to be responsible for the general 30–40 ppb long-term average tropospheric background concentrations (Kelley *et al.*, 1982, 1984), although others consider 10–20 ppb to be more typical summer background concentrations (Altshuller, 1987).

Ozone and other photochemical oxidants can be transported hundreds of miles from areas where precursors originate (US Environmental Protection Agency, 1984). Maximum concentrations develop about 20 miles from the source since scavenging chemicals in urban areas tend to reduce ozone accumulation. Beyond about 20 miles dilution generally becomes important. However, very broad occurrences of elevated ozone concentrations have been shown (Wolff *et al.*, 1982). In one instasnce, a virtual 'ozone river' extended from the Gulf Coast to New England that affected up to a thousand square miles during a one-week period (Wolff and Lioy, 1980). But all this came to light some 30 years after damage to plants from ozone first appeared.

A NEW DISEASE PROBLEM

North America

In the eastern USA, from Ontario to the Connecticut Valley, and south to Maryland, a new disease was causing alarm. 'Weather fleck' of cigar-wrapper tobacco, first observed in 1952 in Beltsville, Maryland, was characterized by straw-colored flecks and lesions, rendering the leaves totally unsuitable for cigars (Wanta *et al.*, 1961; MacDowell *et al.*, 1963) (Figure 8.1). The disease appeared following periods of humid temperature inversions common during the summer and fall months. Hence, the disease was blamed on the 'weather'. But a more explicit explanation was sought. In looking for a causal agent, biologists recognized a similarity of the weather fleck symptoms to those produced when pinto bean plants were exposed to ozone (Middleton

Figure 8.1 Weather fleck of Bel B tobacco. Injury followed a 2-hour fumigation to 100 ppb ozone

et al., 1955). When tobacco was exposed to ozone, white flecks soon appeared on the upper leaf surfaces. Chlorosis and wilting followed. Similar symptoms had been described in a much earlier study (Homan, 1937), but it was only in the 1950s that ozone became recognized as the cause of this and many other serious plant diseases (Darley & Middleton, 1966).

In order to establish whether or not ozone actually was involved in causing weather fleck, it was necsesary to measure ozone concentrations and expose plants to it. A small strip of rubber, bent in half to stretch it and produce a tension, was initially used as a specific test for ozone (Bradley & Haagen-Smit, 1951). The depth of the cracks that appeared provided an index of the intensity of ozone exposures. Methods of wet chemistry also were used, but these were laborious and not always practical in the field. About this time though, a continuous recorder was developed based on color changes produced by the oxidation of iodine. This instrument, the Mast ozone recorder, and others that followed soon replaced the original rubber strips.

Howard Heggestad, with the USDA, and John Middleton at the University of California, two of the pioneers in photochemical pollutant research, were now able to quantify ozone concentrations. They concluded that when ozone concentrations exceeded 200 ppb for at least 3 hours, weather fleck appeared on sensitive tobacco varieties (Heggestad & Middleton, 1959).

Weather fleck was not unique; soon a number of other diseases appeared that were found to be ozone-related. Grape stipple in California was among the first of these (Richards *et al.*, 1958). The disease had been observed since 1954, but the cause remained elusive until grape cuttings were placed in the filtered atmosphere of an enclosed chamber into which ozone was released. The characteristic stipple symptoms appeared on the leaves when vines were fumigated for 3 hours with 500 ppb ozone as measured by buffered potassium iodide. Typical leaf yellowing and leaf drop followed.

The appearance of onion tipburn in Wisconsin by 1954, later established to be caused by ozone, as well as weather fleck of tobacco, confirmed that this pollutant had already become widespread in the eastern USA in the 1950s (Engle *et al.* 1965).

During the late fall of 1958, a New Jersey farmer experienced unusual symptoms on his spinach crop (Daines *et al.*, 1960). Chlorotic to bleached spots appeared on the leaves, making them unmarketable. The Experiment Station staff was asked to diagnose the problem and recommend corrective measures. Inspection and research by Drs Robert Daines, Eileen Brennan and Ida Leone at the Rutgers University in New Jersey (1960) revealed that ozone was responsible. By 1959 the symptoms also had appeared on leaves of many other crops—most notably alfalfa, rye, barley, radish, petunia, beans, broccoli and carnations.

Varieties of some 34 plant species that had been most seriously injured in New Jersey were exposed to varying concentrations of ozone (Hill *et al.*, 1961). Many were visibly injured when subjected to 130 ppb for 2 hours. The possibility that production losses or reduced growth might exist in the absence of visible symptoms remained to be demonstrated.

Symptoms of a similar nature were also being observed in the forest. During the late 1950s the US Forest Service was called upon to look into the general decline of eastern white pine (*Pinus strobus* L.) over an area of several hundred square miles in Tennessee (Berry, 1961; Berry and Hepting, 1964) as well as in West Virginia and North Carolina. Symptoms included premature loss of older needles, yellowing and browning of current year needles, stippling or mottling of foliage, gradual reduction of growth and striking variation in sensitivity among individual trees. In the east the disease was called post-emergence chronic tipburn (Hepting & Berry, 1961). When fumigation studies were conducted, typical symptoms appeared following an exposure of a few hours to as little as 65 ppb ozone (Berry & Ripperton, 1963).

Ozone-induced needle blight of eastern white pine was found to be even more widespread than first suspected. By 1966 it had become prevalent from Ohio to New York and perhaps beyond (Costonis & Sinclair, 1969). Within the next few years needle blight or chlorotic dwarf of eastern white pine was observed throughout its range, from southern Appalachia to Maine. By 1988 symptoms appeared in 23% of the stands sampled in the southern Appalachian Mountains (Anderson *et al.*, 1988).

A similar decline first appeared on ponderosa pines in California during the same period (Parmeter *et al.*, 1962) (Figures 8.2 and 8.3). When first noted in 1953 (Asher, 1956), it was called the 'X' disease and blamed on everything from black pine leaf scale to drought. Viruses, nematodes and fungi were all considered. Later, it became known as chlorotic decline, needle dieback or ozone needle mottle (Richards *et al.*, 1968). The earliest expression was the appearance of minute chlorotic flecks on the older needles. Chlorosis gradually intensified as the leaves became bronzed and finally dropped prematurely. As the decline progressed, needle tip necrosis and shoot dieback developed, and shoot growth was markedly reduced. New needles were stunted and sparse until only a few dwarfed, chlorotic leaves were found at the branch tips. Over a period of a few years the branches continue to die back until the tree is dead. In areas where this decline appeared, daily peak oxidant concentrations often exceeded 100 ppb (Taylor, 1973). Ponderosa pines were moderately to severely damaged if ozone concentration exceeded 80 ppb for 12 hours daily.

Meanwhile, in eastern Canada beginning in 1957, similar symptoms were observed (Linzon, 1960, 1966). Faint pinkish spots on the stomata developed into orange-red bands that spread to the needle tips within a few days. Since the symptoms were initiated only in semi-mature leaf tissue, this disorder was named semi-mature needle blight or SNB. Each major outbreak appeared following a day or more of quiet weather. However, there were some distinct differences between SNB and the ozone-induced emergence tipburn.

Europe

By the 1960s the presence of ozone was also being questioned in Europe, and the sensitive Bel W_3 tobacco variety was used to monitor ozone throughout the UK (Bell & Cox, 1975; Ashmore *et al.*, 1978). Injury was found to be widely distributed although most severe in Wales and central Scotland. Studies in Western Europe largely between 1973 and 1980 revealed that foliar necrosis on Bel W_3 correlated closely with the number of hours ozone concentrations equalled or exceeded 40 ppb (Bell, 1984). Maximum

Figure 8.2 Chlorotic decline of pine, showing the difference in susceptibility among individuals. (Photo courtesy of Paul Miller.)

1-hour concentrations up to 180 ppb were reached in rural areas in 1975 and up to 270 ppb in Vlaardingen and 258 ppb near London in 1976 (Becker *et al.*, 1985). Concentrations in other areas of Europe were slightly lower during this period. Since 1979 ozone concentrations in much of Europe, from Norway to Greece, have exceeded 200 ppb (Becker *et al.*, 1985).

SPECIES SENSITIVITY

The eastern white pine would seem to be in a class by itself as to ozone sensitivity, with ponderosa pine in the west not far behind (Costonis &

Figure 8.3 Chlorotic stipple of ponderosa pine needles produced by ozone fumigation

Sinclair, 1969). But what of the other conifers, about which less is written? Observations in the forest have been supplemented by controlled fumigation studies in chambers, but chambers have the limitation that the sensitivity of the limited seedling populations used may not be representative of the broader population. Another problem lies in the exposure dynamics.

In California field observations showed that Jeffrey pines generally had about the same sensitivity to ozone as ponderosa pine, with white fir less sensitive and incense cedar and sugar pine more tolerant, although all developed chlorotic symptoms in areas where ponderosa pines were severely damaged (Miller, 1973). Fumigation studies that evaluated the sensitivity of other species showed western white pine and red fir to be as sensitive as ponderosa pine. Coulter pine, Douglas fir, Jeffrey pine, white fir, big-cone Douglas fir and knobcone pine were more tolerant; while incense cedar, sugar pine and giant sequoia were most tolerant.

At the Pennsylvania State University, another early center for air pollution studies, Drs Don Davis and Al Wood (1972) exposed 18 species of 2- to 6-year-old coniferous tree seedlings to 100 ppb ozone for 8 hours or to 250 ppb for 4–8 hours. In this study, Virginia pine, jack pine, European larch, Austrian pine and Scotch pine were all more sensitive than eastern white pine, based on the percentage of the fumigated population that showed injury and the severity of the injury. Many more species were tested for ozone sensitivity during the ensuing years, and in 1976 Davis and Wilhour published a compilation of the findings (Table 8.1). The notable inconsistencies apparent in this table, such as the intermediate sensitivity of eastern white pine and the tolerance of white fir, largely are attributed to differences among clones. Also, inferences cannot be drawn as to the possible subliminal effects ozone may have on these species in the absence of clearly visible symptoms.

Despite the extreme sensitivity of some conifers to ozone, the overall sensitivity of woody species is no greater than that of herbaceous species. Some cultivars are particularly sensitive, such as the Bel W_3 tobacco variety. Differences in sensitivity among varieties of many species make it difficult, if not unrealistic, to compare species sensitivity. Nevertheless, a knowledge of relative sensitivities is valuable in diagnosing potential air pollution effects.

When we consider the differences in sensitivity among species, we tend to ignore, or pass off lightly, the genetic differences among varieties, cultivars and even individuals of a single species. Yet these may be far greater than species differences.

Researchers have attempted to determine the genetic basis for such differences. In the case of soya beans the ozone injury response may be controlled by a few genes. Based on the lack of dominance and injury response distribution in the F_1 and F_2 populations, partial dominance and gene interaction may be involved. A two-gene model with complete dominance of one locus and partial dominance and epistasis of the other was suggested. Back-crosses supported a two-gene model.

Ozone tolerance in bean and onion is qualitatively inherited by a few genes and one gene, respectively (Engle & Gabelman, 1966). Additional

Table 8.1. Relative sensitivity of woody plants to ozone[1]

Sensitive	Intermediate	Less sensitive[2]
Ailanthus	Ash, European mt.	Arborvitae
Ash, green	Elm, Chinese	Azalea
Ash, white	Forsythia	Birch, European
Azalea	Gum, sweet	Dogwood, gray
Cotoneaster	Hemlock, Eastern	Dogwood, white
Honey locust	Larch, Japanese	*Euonymus alatus*
Larch, European	Mock orange	Fir, balsam
Oak, white	Oak, pin	Fir, Douglas
Pine, Austrian	Oak, scarlet	Fir, white
Pine, Jack	Pine, eastern white	Firethorn
Pine, ponderosa	Pine, pitch	Gum, black
Pine, Virginia	Pine, Scotch	Holly, American
Poplar	Redbud, eastern	Laurel, mountain
Sycamore, American	Rhododendron	Linden, American
Maple, Norway	Viburnum	Linden, little leaf
		Maple, sugar
		Oak, English
		Pine, red
		Privet
		Rhododendron, Carolina
		Spruce, Black Hills
		Spruce, Colorado
		Spruce, Norway
		Spruce, white
		Yew

[1] Adapted from Davis & Wilhour (1976) and based on the appearance of visible symptoms. However, the research on many of these species is incomplete and subject to modification pending further research.
[2] The term 'less sensitive' is used here as advocated by Guderian et al. (1985), rather than tolerant, so as not to infer that plants can be completely resistant to ozone.

genes may be involved with tobacco, tall fescue, sweet corn, petunia and potato. This indicates that the type of inheritance may vary with the crop.

Many of the sensitivity studies have involved short-term fumigations and the appearance of visible symptoms. This provides a basis for designating categories of sensitivity, as summarized in Table 8.2. But it is not the whole story. Although exposures to relatively high concentrations of ozone for an hour or two can account for most of the visible symptoms, exposures to lower concentrations for longer periods are likely to account for most losses to production despite the absence of visible symptoms. Therefore, it is not entirely feasible to list a threshold concentration at which injury first appears

Table 8.2. Relative sensitivity of agricultural crops and weeds to ozone[1]

Sensitive	Intermediate	Less sensitive[2]
Alfalfa	Cabbage	Beet
Barley	Carrot	Cotton
Bean	Corn, field	*Descurainia*
Buckwheat	Cowpea	Jerusalem cherry
Citrus	Cucumber	Lamb's-quarters
Clover, red	Endive	Lettuce
Corn, sweet	Hypericum	Mint
Grape	Parsley	Piggy-back plant
Grass, bent	Parsnip	Rice
Grass, brome	Pea	Strawberry
Grass, crab	Peanut	Sweet potato
Grass, orchard	Pepper	
Muskmelon	Sorghum	
Oat	Timothy	
Onion	Turnip	
Potato		
Radish		
Ragweed		
Rye		
Safflower		
Soybean		
Spinach		
Tobacco		
Tomato		
Wheat		

[1] Adapted from Lacasse & Treshow (1976).
[2] The term 'less sensitive' is used here as advocated by Guderian *et al.* (1985), rather than tolerant, so as not to infer that plants can be completely resistant to ozone.

and delimit sensitivity categories by a single numerical value. Sensitivity categories thus are highly generalized.

SYMPTOMATOLOGY

General

Although the specific symptoms of ozone injury vary among plant species and varieties, certain general expressions form a common thread of similarity. The reasons for this lie in the fundamental mechanisms of toxicity.

Several factors must be considered when attempting to determine if ozone is the cause of a particular injury. As with any diagnosis, the observer must be able to recognize symptoms caused by as many stresses as possible. Those caused by known biotic pathogens such as fungi and insects, as well as abiotic stresses related to adverse temperature, moisture or soil relations, must be recognized.

The injury caused by ozone is first characterized by the appearance of minute chlorotic to pale tan or whitish lesions on the upper surface of the affected leaf. The lesions, no more than flecks, initially are less than a millimeter across, but, especially should they coalesce, can become much larger. While initially limited to the upper leaf surface, lesions may extend through the leaf when ozone concentrations are higher. The early flecks typically are restricted between the smaller veins but overlap these as the injury progresses.

The degree of leaf maturation also influences responses. Ozone first affects the base of the leaves; injury then develops progressively towards the tip on successive days as the leaf matures. Next the bases of the younger leaves are affected. Thus on a plant such as tobacco, where leaves clearly mature from the lower to upper stem, it is easy to trace the progress of the ozone symptoms and even the days when exposure occurred.

Specific symptoms may vary among different kinds of plants. Thus we see some differences largely between herbaceous plants, woody plants and the grasses and cereals. Needle-leaved conifers also show some distinct patterns (Lacasse & Treshow, 1976; Hill *et al.*, 1970).

Herbaceous plants

Upper leaf surface stipple is particularly characteristic on herbaceous plants, especially those having a well-developed palisade tissue. Initially only a few cells are involved, but as additional cells become affected the lesion enlarges to become visible to the naked eye. As the affected cells collapse, the tissue typically assumes a light tan or grayish to milky-white appearance. Lesions may reach a size of a few millimeters or even a few centimeters across, if enough lesions coalesce. Lesions tend to be irregular in outline, being roughly restricted by the smaller veins. The spongy parenchyma and epidermis are not usually injured unless ozone concentrations are very high. Lesions on leaves of some species become noticeably pigmented with flecks ranging in color from light to dark brown or reddish. Soya bean provides one example, dock another, where excessive amounts of the red anthocyanin pigments are formed. The fleck-type lesions still occur but they are distinctly reddish in color (Hill *et al.*, 1970; Lacasse & Treshow, 1976). Anthocyanins also may be formed in response to other stresses such as low temperatures or physical injury. Reddening may extend beyond the fleck lesions and encompass much of the leaf.

Occasionally, as in spinach, one observes a general shiny or waxy symptom on the upper leaf surface. This may disappear in a few hours if the exposure ceases, or may develop to produce a dull, orange-green color during a subsequent fumigation. Damaged cells may survive complete death and desiccation, but in such instances the chloroplasts are at least partially disrupted and the amounts of chlorophyll are noticeably reduced.

Within this broad context, specific symptoms may vary among species. On spinach, for instance, lesions are characteristically large and bleached, while on alfalfa the symptoms more often consist of minute whitish to yellow-green punctate flecks (Hill *et al.*, 1970). In other instances light-green to yellowish chlorosis may extend over much of the leaf, with sharply defined small islands of normal green tissue scattered within the chlorotic tissue.

The best-known and most 'classic' symptom is exemplified on tobacco. Numerous small lesions appear mostly on the upper surface of fully expanded leaves, although they are often bifacial on the most sensitive varieties. Wilting follows when ozone concentrations are higher.

Grasses and cereals

Tiny, chlorotic or white to tan flecks also best characterize symptoms on cereals and grasses. The initial flecks frequently coalesce between the larger veins to form chlorotic to bleached streaks or oblong lesions. Since a palisade cell layer is lacking in cereals, the injury extends through the leaf. Injury is usually most intense at the apex of a bend, with necrosis most prominent at the leaf tips and margin as with onion tip burn. Bifacial necrotic streaks develop on sweet corn. The individual irregularly shaped streaks are relatively large, concentrated along the margins, and frequently include the larger veins as on corn (Figure 8.4). Younger plants are most sensitive.

Woody, broad-leaved plants

Injury to shrub and tree foliage, while essentially similar to that on herbaceous plants, tends to be more limited to the upper leaf surface than on annual crops. Epidermal cells remain normal in appearance. The punctate lesions tend to be more brownish and bounded by the smallest veins (Figure 8.5). Their abundance when dense, however, can give the affected leaf a bronzed appearance as the lesions coalesce, although color may range from silvery to purplish depending on the plant species. In some plants, such as aspen, the intercostal lesions have a blackish, blotched appearance (Figure 8.6). Unpigmented 'bleaching' is less common than on herbaceous plants. The dark, punctate symptoms of grape stipple, among the earliest and best known of the ozone-involved disorders, are typical of those on woody plants. Subsequent bronzing and premature senescence contribute to the production losses attendant with the disease. Symptoms on many wood species are virtually indistinguishable from early senescence, and the physiological

Figure 8.4 Necrosis on corn following one, $2\frac{1}{2}$-hour exposure to 470 ppb ozone

mechanisms in both cases may be similar. Defoliation in midsummer has long been recognized as a response to smog in the Los Angeles area.

Gymnosperms

The varied descriptive names of conifer diseases attributed to ozone reflect the different symptoms that ozone can induce. White pine needle blight, chlorotic dwarf, chlorotic fleck, chlorotic mottle and emergence tipburn are all accurately descriptive. The general symptoms on pines are similar, but there are specific distinctions and variations, apparently depending as much

Figure 8.5 Ozone stipple lesions on the upper surface of an avocado leaf

on the individual genetic tree variation, and the conditions of ozone exposure, as on the species (Dochinger & Seliskar, 1970; Costonis & Sinclair, 1969; Parmeter *et al.*, 1968).

The chlorotic dwarf symptom may reflect the extreme sensitivity of certain individuals as much as anything. Affected trees are characterized by 'stunted roots and tops, short, mottled needles, and premature shedding of foliage' (Dochinger & Seliskar, 1970). Current-year needles may be thin and twisted, sometimes showing necrotic tips, and older foliage is usually shed before the new needles mature. When charcoal-filtered chambers were placed over such affected trees for several years, the new foliage and growth was healthy. When these chambers were subsequently removed, growth became retarded and symptoms of chlorotic dwarf, mottling and needle drop once again appeared. In this study symptoms were attributed in some instances to the presence of sulfur dioxide in combination with ozone.

Symptoms of ozone injury on conifers essentially involve flecking, although needle tip necrosis occurs less frequently. Specifically, the initial macroscopic symptoms of ozone injury appear as minute silver flecks radiating from the stomata of current-year needles. These flecks develop into larger, diffuse-margined, chlorotic flecks visible to the naked eye. Semi-mature tissue is

Figure 8.6 Lesions characterizing ozone injury on quaking aspen (4-hour ozone exposure at 250 ppb)

the most seriously affected portion of the needle, but mature and immature tissues may be affected simultaneously. Pink lesions and bands may develop from these lesions, spreading and extending to the needle tips.

In many cases injury does not progress beyond the flecking stage. The reasons for this are not entirely clear, although sensitivity of the individual trees is considered to be important (Hill *et al.*, 1970). Environmental parameters, such as temperature and moisture relations during exposure, may also play a role in the symptom characteristics (Costonis & Sinclair, 1969).

All needles of a fascicle are not equally affected by an exposure to ozone and some may escape injury. Symptom severity also may vary among fascicles and twigs. Variation among trees is even greater owing to differences in genetic sensitivity to ozone. One tree may have 70% of the total current-year needle tissue necrotic, while an adjacent tree of the same species remains completely free from symptoms. Needles of the most susceptible trees are shorter than normal, chlorotic, and have conspicuous tip necrosis. One-year-old needles often drop by midsummer. Slightly more tolerant trees have more normal green color and show only the yellow to brown lesions associated with the stomata. Needles on such trees may be shed one to two years prematurely. This is the culmination of the premature senescence and ultimate chlorotic dwarf caused by ozone.

In California injury symptoms also have been described on white fir and sugar pine. Less definitive chlorotic mottle has been observed on giant sequoias, incense cedar and lodgepole pine (Williams *et al.*, 1977). South of Mexico City symptoms and decline are severe on the sacred fir (*Abies religiosa*) (Figure 8.7).

Mimicking symptoms

Several environmental stresses can cause symptoms that can be mistaken for ozone injury (e.g. Treshow, 1970; Malhotra & Blauel, 1980; Hill *et al.*, 1970). The most similar of these are caused by physiological stresses, including other air pollutants; but viruses, bacteria, mites and insects also can cause mimicking symptoms. Many of these, however, can be distinguished by the appropriate diagnostic procedures.

The greatest problems come with diagnosing the various necrotic and chlorotic markings on conifers, most notably pines. A disease of eastern white pine, generally known as 'needle blight', has been especially troublesome to diagnose. The disease, found throughout the northeastern USA, is characterized by stunting, chlorosis and/or tip necrosis of needles (Dana, 1908). The detailed description of symptoms in the original paper is identical to symptoms now often attributed to ozone. In 1908 the disease affected about 10% of the white pine trees—the same as found in a 1984 survey (Treshow). But could ozone have been present in toxic concentrations in the early 1900s when the disease was first described?

A second disease, semi-mature tissue needle blight (SNB), has caused further complications in diagnosis by its similarity to ozone injury. Differences noted are that SNB, as the name implies, develops only before the needles become mature. Also, microscopically, SNB is initiated only in mesophyll cells that are in the stage of needle tissue maturation where suberization of the endodermal cells is proceeding. With ozone injury, mesophyll cells collapse simultaneously in different stages of needle tissue maturation, although injury is most pronounced in semi-mature tissue. SNB develops only during the 9–10-week period during which needles are maturing, while ozone injury occurs following ozone episodes. Weather seems to be involved with SNB since the injury occurs only when wet periods are followed suddenly by dry periods.

Figure 8.7 Decline of sacred fir (*Abies religiosa*) forests south and southwest of Mexico City, known as Desierto de los Leones. (Photo courtesy of G. Elizalea and M. L. de Bauer.)

Macroscopically, SNB appears first on stoma-bearing surfaces as faint pinking spots in semi-mature tissue. These progress into orange-red bands that spread through more mature tissue towards the needle tip. Ozone injury typically involves more general flecking in addition to the tip burn. Some of the arguments are relatively obscure, and there is still controversy as to whether or not SNB might be caused by ozone. But almost no research on SNB has been done in the past two decades (Bennett *et al.*, 1986).

Another flecking symptom, similar to that caused by ozone, is known as winter fleck (Miller & Evans, 1974). The cause is not known but is associated with winter weather conditions, especially exposure to snow. Lesions of winter fleck tend to be smaller and more sharply delimited than ozone-induced lesions, more randomly scattered, and limited to the abaxial surfaces exposed to the sky. Histologically, hyperplasia in the endodermis and transfusion area are common only with winter fleck.

Low-temperature stress symptoms also can resemble ozone injury on broad-leaved species. Radiation frost, especially, can produce white flecks on leaves of alfalfa, spinach and other crops that resemble the effects of ozone. Such responses would only pose a diagnostic problem if frosts were concurrent with an ozone episode, which is unlikely.

Adverse moisture relations may also complicate diagnosis. This can be especially true where flooding has occurred, even for only several hours (Treshow, 1970). The white spot disease of alfalfa, clover and similar species is characterized by sharply delimited necrotic spots usually only a few millimeters in diameter. It develops when plants growing under fairly dry conditions are suddenly exposed to ample or excess moisture over a period of a day or two. The disease is aggravated if above-normal temperatures prevail.

Chlorotic mottle of pine, as well as flecking on broad-leaved plants, has also been associated with nutrient deficiencies (LaCasse & Treshow, 1976). Deficiency of potassium may cause the development of bleached or brown spots, but these are generally located along the margins, and are associated with leaf curling. Such yellow or brown spotting is most distinctive on legumes, notably alfalfa and cotton (Treshow, 1976).

Whitish stipple, together with necrosis indistinguishable from that caused by ozone, also can be caused by chlorine. Although there are subtle histological differences (Treshow, 1970), the most practical approach to diagnosis lies in the differences in relative sensitivity among plant species. Also, chlorine injury is rare and is associated with point sources. Hence injury is local in distribution rather than being dispersed over a broad, regional area as with ozone.

The biotic pathogens causing symptoms most readily mistaken for ozone injury on both conifers and broad-leaved species are mites and leaf hoppers. Their sucking feeding habit leaves minute whitish flecks that can mimic ozone symptoms, as can the bronzing or silver glaze produced by eriophyd mites. But careful examination of the leaf will generally reveal the insects, the remains of their dead bodies, or webbing in the case of some mite species (Treshow, 1970).

SUGGESTED READING

Cooley, D.R., & Manning, W.J. (1987). The impact of ozone on assimilate partitioning in plants: a review. *Environ. Pollut.*, **47**, 95–105.

Zerefos, C.S., & Ghazi, A., eds (1985). *Atmospheric Ozone.* Published for the European Communities by Reidel, Boston, Massachusetts. From a symposium in Halkidiki, Greece, September 1984. xxxii + 842 pp., illus.

CHAPTER 9

Ozone Research and Discovery Come of Age

RESEARCH APPROACHES

Exposure chambers

It seemed perfectly rational at the time. Place groups of similar plants in each of two chambers; add exhaust fumes and ozone to one of the chambers, and let the other serve as a control; measure growth differences between the plants in each chamber. The year was 1952; the place was Riverside, California, 60 miles east of Los Angeles. The air was clear and was presumed to be free of pollution. After several weeks of finding no differences in growth, scientists realized the air was not clean, not in the control chamber and not in the ambient atmosphere. Photochemical pollution, known then only as smog, had already arrived in Riverside in toxic concentrations.

One major lesson was learned from this pioneer work: it was necessary to filter any air used in an air pollution study. A second lesson came much later: it was necessary to use more than two chambers. Adequate replication for statistical analysis required more than the numerous plants in each chamber. It demanded replications of the chambers for each treatment (Oshima & Bennett, 1978; Sokal & Rohlf, 1969).

Two main approaches for studying the responses of plants to pollutants, including ozone, were used. Primarily, known concentrations of a pollutant were added to chambers that were in themselves within a greenhouse. A second approach subjected plants in the separate chambers to ambient air as well as known pollutants. In both cases growth parameters were compared between plants in chambers having polluted air and plants grown in filtered air. Activated charcoal filters have been routinely used for many years, but now it seems that cardboard tubes are even more effective. While use of

105

ambient air is realistic in terms of the pollutants to which plants are actually exposed in the field, it does not isolate or quantify the specific toxic components in the air. Thus, over the years, regular fumigations with known amounts of a given pollutant became preferred, although ambient air was often included as one of the treatments (Heck *et al.*, 1978).

In order to better quantify plant responses to ozone, plants are grown in chambers to which known amounts of ozone are introduced. Tank oxygen or air is passed over an electric arc to produce ozone. However, when air is used, nitric acid (HNO_3) and its anhydride dinitrogen pentoxide (N_2O_5) are produced, which may also affect the plants being studied. The source of the nitric acid and anhydride that appears with ozone in the air stream passed through an electric arc is atmospheric nitrogen dioxide (NO_2). One method of producing nitric acid (and its anhydride, dinitrogen pentoxide) is to oxidize nitrogen dioxide with ozone. Nitrogen pentoxide is unstable and decomposes spontaneously into nitrogen dioxide and oxygen. The most stable oxides of nitrogen are nitrogen dioxide and nitric oxide.

Scientists also have been concerned that plants grown in chambers might not respond the same as plants grown under natural field conditions. Several approaches have been used over the years in efforts to circumvent this problem and achieve more natural growing conditions. These range from having closed chambers with a rigidly controlled environment to open-air fumigations in which pollutants were released into the atmosphere from conduits to drift over plants grown in adjacent rows (Heagle & Philbeck, 1978). All such approaches provided valuable information so long as certain requisites were met. The essential characteristics of chambers are that they allow a uniform concentration of the pollutant to circulate throughout the chamber and between chambers, provide a uniform environment, have a non-reactive surface, permit precise control over pollutant concentrations and have an environment resembling ambient conditions.

Open-top field chambers

While several chamber designs are in use that meet these requisites, the one currently preferred is an 'open-top' design in which an upward positive air flow through the chamber minimizes the incursion of ambient air through the open top of the chamber while allowing more natural light and moisture conditions to prevail in the chamber (Mandl *et al.*, 1973; Heagle & Philbeck, 1978). A cylindrical shape facilitates uniform distribution of the pollutant when it is released from a large tube around the base of the chamber. The main disadvantage is the ingress of ambient air during periods of strong wind. However, a baffle at the top of the chamber largely prevents the downward turbulence that could allow mixing. Open-top chambers were utilized in the US EPA-sponsored National Crop Loss Assessment Network

(NCLAN), in which the impact of ozone on production was studied in major US agricultural regions.

Field exposures

Throughout the history of air pollution biology, the classic approach to studying plant responses has been to observe plants close to a pollutant source and compare their injury symptoms, if any, and growth characteristics with plants remote from the source or along a gradient of concentrations. While this has been a valuable technique, it was difficult to quantify differences because growing conditions and environments were not the same over the transect (Oshima *et al.*, 1976). This was especially true in agricultural situations.

With forests, comparisons of growth could be better quantified. This is especially true when dendrochronological techniques are utilized to study radial increment growth of stands selected along one or more transects from a known pollutant source. This works well with a point source, but becomes more difficult when studying responses to a regional pollutant such as ozone.

EXPOSURE DYNAMICS

The response to ozone, whether visible injury or the subtle subliminal effects are addressed, depends on more than the average concentration. It depends on the duration of exposure. More than this, responses also depend on: (1) the brief, peak concentrations when the exposure took place (both daily and seasonal peaks); (2) the intervals between exposure; and (3) the environmental conditions at the time of exposure and between exposures. The number of hours that ozone concentrations exceed a given value may be most important of all. Such potential exposure combinations are referred to as 'exposure statistics' or 'exposure dynamics'.

Ozone research has always attempted to simulate natural, ambient concentrations and exposure conditions. But which exposure regime does one choose as being representative? The possibilities are infinite, and it is impossible to study all the potential regimes. All we can do is attempt to pick a surrogate value that best reflects what is happening in the field (Rawlings *et al.*, 1988). Ambient ozone concentrations in the summer typically range between 20 and 200 ppb. A background concentrations is often considered to be 25 ppb but can briefly reach 40 ppb in some situations. At higher elevations, prolonged concentrations of 30–40 ppb are common. Early research utilized peak concentrations of roughly 150–300 ppb for a few hours in an attempt to produce visible symptoms and learn something about the relative sensitivity of species (e.g. Hill *et al.*, 1961). More recent research has attempted to learn what effects ozone might have on production at the more common, lower concentrations and longer exposure periods.

There are many ways in which exposure dynamics have been expressed. The maximum hourly daily mean is one, but it does not provide the total duration of exposure. Another value, the average 24-hour mean, fails to give the peak concentration. Other exposure statistics that have been considered include 7-hour. 12-hour, seasonal means, and hourly concentrations above a given concentration (e.g. above 70 or 100 ppb, etc.) (Lefohn & Jones, 1986; Pinkerton & Lefohn, 1987).

The summary values reported by the State Monitoring Networks in the US are largely limited to the hourly averages and the number of exceedences of the Air Quality Standard, currently 0.12 ppm (120 ppb) and second-highest concentration. These data do not provide a suitable value on which to base the impact of ozone. They are more suitable in providing a rough idea of air quality and comparing trends in concentrations over time and among areas.

The most valid way of determining the exposure statistic that is best correlated with yields is obtained when ample yield data are available under known exposure conditions. This comparison was possible when all the NCLAN data were analyzed over a two-year period (Lee *et al.*, 1987). The top-performing exposure indices proved to be those that: (1) cumulated the hourly concentration over time; (2) used a sigmoid weighting scheme that emphasized concentrations of 60 ppb and higher; and (3) phenologically weighted the exposure such that the greatest emphasis was placed on the most sensitive plant growth stage.

These principles were supported in a reanalysis of the NCLAN wheat and soya bean data. A peak-weight statistic and the number of occurrences of over 80 ppb, or the sum of the concentrations over 80 ppb, had a greater correlation with actual yields than a 7-hour statistic (Lefohn *et al.*, 1988).

The peak concentrations, or even concentrations above a certain value, are important, at least in part, because of their role in causing the stomata to close, thus reducing carbon dioxide uptake and photosynthesis. The stomatal interaction becomes most serious when the time between succeeding high concentrations is inadequate for recovery.

Ozone concentrations, notably the second-highest hourly concentrations, are summarized annually by the US EPA (e.g. US Environmental Protection Agency, 1984). Highest concentrations are found in the northeast, the Gulf coast and the west coast, where in 1983 the second-highest daily maximum 1-hour concentrations averaged 160, 170 and 210 ppb, respectively. The second-highest concentrations in the individual cities were 250 ppb in Newark, New Jersey; 280 ppb in Houston, Texas; and 370 ppb in Los Angeles, California.

Ozone concentrations in more industrialized parts of Europe approach this same general range but tend to be somewhat lower owing both to the higher latitude and more moist weather conditions.

SUBLIMINAL EFFECTS OF OZONE

Researchers have considered that possible 'hidden' effects of air pollutants, including ozone, may far outweigh the importance of the obviously visible symptoms of the disorders caused by pollutants. Originally the term 'hidden' simply referred to effects that could not be detected visually. This would largely involve losses in growth or production, but also includes reduction in quality or other impairments of a crop. Technically, such effects would not be 'hidden' if they existed and could be measured. Improved technology and instrumentation have made it ever more possible to measure such subtle changes. Although scientists in the air pollution field know what is inferred by the term 'hidden injury', alternative terms have been sought. The term 'latent', popular in Europe, comes close but infers that the effect has yet to develop. The word 'subliminal' more accurately denotes effects that already exist but are not readily apparent.

The presence of subliminal effects was controversial for many years, but research over the past 20–30 years has clearly established that such adverse physiological responses not only occur for ozone, but represent the greatest threat to plant growth and production.

Establishing thresholds

What concentrations of ozone can cause adverse effects, and how long an exposure to such concentrations is needed? Establishing this dose–response relationship has been a major goal of air pollution biologists from the beginning. The exposure regimes used are vital to estimating yield losses and economic effects of ozone to agriculture, and they are critical in evaluating forest and other plant community responses.

The question of which exposure characteristics to use has been asked again and again over the past 30 years. In seeking answers the initial emphasis was on the concentrations required to cause visible injury. But over time the significance of subliminal effects in the absence of visible symptoms directed concern towards determining the lowest exposure at which growth and production are first impaired.

The principal approach used to determine thresholds has been to fumigate plants grown in chambers or greenhouses under controlled environmental conditions and then expose the plants to known concentrations of ozone for prescribed periods of time. A second approach, sometimes referred to as 'reverse fumigation', is to expose plants in one group or set of chambers to ambient, polluted, atmospheres but filter the air going into the second set of chambers. Data in both cases are examined to learn the lowest concentrations at which some effect is measurable and the duration of exposure required for it to appear. These thresholds can be extremely variable, not only among species but even among cultivars, because of genetic differences as well as different fumigation methods and environmental conditions.

Production losses

Production loss, or yield reduction, refers to more than yield. The concept includes any impairment of the intended use of the plant: loss in weight, number or size of plant parts that might be harvested; changes in chemical composition or quality; or loss in esthetic quality, a value difficult to quantify or even judge. Focus has been mostly on reductions in weight of the marketable plant organ, so this provides the major basis for assessing losses and determining the threshold at which losses become significant.

Considerable research in the 1960s and 1970s showed that ozone adversely affects production. This was reviewed by Taylor (1984) and, in greater depth, in the 1986 US EPA Air Quality Criteria Document. Notable early studies (Heagle, 1972; Heagle *et al.*, 1974, 1979, 1980) showed that low concentrations of ozone could cause pronounced losses in production when exposures were extended through the growing season. Ozone concentrations as low as 50 and 100 ppb for 6 hours per day throughout the growing season caused significant reductions in fresh weight of corn ears, number of kernels and dry weight of kernels. Seasonal 7-hour per day exposures to 100 and 150 ppb significantly reduced yields of spinach, winter wheat, field corn and soya bean.

In some plants, root growth was found to be reduced even more than top growth. This may be attributable to the inhibition of starch translocation to the roots caused by ozone (Hanson & Stewart, 1970). A secondary effect may be in reducing populations of the root-inhabiting mycorrhizae (McCool *et al.*, 1979).

These and other excellent studies in the 1970s gave a general picture of the harm that ozone might do, but they did not provide a full view of the impact of ozone on national or global yield reduction. In order to address these questions, the National Crop Loss Assessment Network (NCLAN) was established by the US EPA in 1980 (Heck *et al.*, 1982). This program was integrated among several research institutes located in major agricultural areas of the USA. Standardized methodologies were used, and many of the same crops and cultivars were grown in the different regions.

Two major criteria helped determine the plant species that were selected for study: sensitivity to ozone based on earlier research, and the importance to US agriculture. Species studied included soya bean, alfalfa, wheat, cotton, peanuts, tobacco, and the forage crops clover and fescue.

The cultural conditions used approximated typical agronomic practices. A range of ozone concentrations was studied, with sufficient replicates to develop exposure–response models from which economic losses could be determined. Ozone concentrations used were based on those typically occurring in ambient air, and open-top field chambers were supplied with ambient air or air supplemented with ozone to provide concentrations three or four levels greater. Charcoal-filtered chambers provided controls. Ozone was added to the chambers 7 hours daily, and exposures typically were characterized by a 7-hour seasonal mean ozone concentration.

The relationship between ozone concentration and plant yield was based on regression equations that predicted the loss that would result from specific ozone concentrations. Wheat, kidney bean and Hodgson soya bean were most sensitive. Ozone concentrations of 28–33 ppb at a 7-hour seasonal mean were predicted to cause a 10% yield loss. At 40 ppb, yield reductions ranged from none in sorghum, barley and a corn cultivar, to as much as a 28.8% loss in Vona wheat.

Approximately 57% of the 37 species of cultivars were predicted to have a 10% yield loss at 7-hour mean ozone concentrations below 50 ppb. Another 35% would experience a 10% loss between a 40–50 ppb mean. Mean concentrations in excess of 80 ppb were required to cause a 10% loss in 19% of the cultivars studied (Table 9.1).

The NCLAN study showed that several cultivars were seriously affected at ozone concentrations no higher than what is commonly regarded as 'background'. It also showed that variation in sensitivity among cultivars can be greater than among species (US Environmental Protection Agency, 1986).

In addition to the 7-hour mean seasonal average ozone concentration initially studied, various other regression equations were tested to determine if any fit the yield data better. Plateau-linear or polynomial equations proved superior to any linear equations. The Weibull model regression slope produced a curvilinear response line that provided a reasonable fit to most data (Heck *et al.*, 1984). It must also be noted that the best fit between an exposure statistic and yield varies even among varieties of species as well as among species.

When 12-hour seasonal means are compared with 7-hour seasonal means, it appears that the 12-hour means shows a still better correlation with yield responses, at least with tobacco and soya bean (Heagle *et al.*, 1987).

Other research suggests that a 24-hour average of the ozone concentrations provides a better fit with the exposure–response data than either 7- or 12-hour averages since crop yield reductions result from an accumulation of

Table 9.1. Summary of ozone concentrations predicted to cause yield losses[1]

Species	Percentage loss predicted to occur at 7-hour seasonal mean ozone concentration of:	
	40 ppb	60 ppb
Legume crops		
Soya bean, Corsoy	6.4	16.6
Soya bean, Davis (81)	11.5	24.1
Soya bean, Davis (CA 82)	6.4	16.5
Soya bean, Davis (PA 82)	2.0	10.4
Soya bean, Essex	7.2	14.3
Soya bean, Forrest	1.7	5.3
Soya bean, Williams	10.4	18.1
Soya bean, Hodgson	15.4	18.4
Bean, kidney	14.9	28.0
Peanut, NC-6	6.4	19.4
Grain crops		
Wheat, Abe	3.3	10.4
Wheat, Arthur 71	4.1	11.7
Wheat, Roland	10.3	24.5
Wheat, Vona	28.8	51.2
Wheat, Blueboy II	0.5	2.8
Wheat, Coker 47-27	2.2	8.4
Wheat, Holly	0.0	0.9
Wheat, Oasis	0.4	2.4
Corn, PAG 397	0.3	1.5
Corn, Pioneer 3780	1.4	5.1
Corn, Coker 16	0.0	0.3
Sorghum, DeKalb-28	0.0	2.7
Barley, Poco	0.0	0.5

[1] Based on Heck *et al.* (1984).

daily ozone effects over the growing season (Cure *et al.*, 1986). However, the fact that means treat high and low values as having the same impact, and minimize the contribution of peak concentrations, would suggest that totalling the concentrations higher than a given concentration might be preferable.

This exposure statistic—the number of hours each day, over a period of days, to which plants are exposed to ozone—also can be used to develop thresholds. This gives the total cumulative number of hours of exposure (Posthumus, 1985). Naturally some problems arise with this simplistic approach: under natural, ambient conditions, peak concentrations of shorter duration may be important, and these are not simulated in the controlled exposures. Intervals between exposures that provide a period of potential recovery also may be important, and these are not expressed in the cumulative dose. Nevertheless, this method gives an approximation that is consistent with other summaries (Heck *et al.*, 1982; Jacobson, 1977; Guderian *et al.*, 1985).

Hourly concentrations above a given value provide some of the best correlations with yield losses (Adamait *et al.*, 1987). The strength of the correlation is influenced by the crop sensitivity. In the case of the moderately sensitive white bean, highly significant correlations were found when ozone concentrations above 80 ppm were calculated and rain and temperature data were incorporated.

Analysis of variance tests also were applied to the data. The lowest ozone concentration that significantly reduced yield was determined; this was often the lowest concentration used. Consequently, it was not always possible to estimate a no-effect concentration. Generally though, concentrations of 100 ppb for a few hours per day for several days to several weeks caused significant, 10–50%, yield reductions.

The California Air Resources Board (1987) has compared crop losses in the field under ambient conditions with loss estimates based on NCLAN results. The CARB used 12-hour growing season average ozone concentrations. A 12-hour growing season average of 60 ppb had approximately the same impact on crop yields as ambient air quality for a representative year, 1984 (Table 9.2). If the average ozone concentration was 50 ppb, yield losses would be 75% of those exposed to 60 ppb. They estimated that even a 12-hour average of 40 ppb would cause significant, but lesser, losses.

Another approach for determining yield losses is to treat randomized test plants growing in areas of high ambient ozone with a chemical that eliminates any effect of ozone. Matching plots are left untreated to serve as controls. The chemical most effectively used to mitigate the effects of ozone, and having minimal to no side effects, is ethylene diurea (EDU) (Manning *et al.*, 1974; Taylor & Rich, 1974). However, recent data indicate that EDU has side effects toxic to the plant.

Despite the question of the phytotoxicity of EDU, studies with the chemical in ozone-polluted areas has supported the NCLAN models in several cases. EDU applied to field plantings of onion in eastern Canada increased yields as much as 37% (Wukasch & Hofstra, 1977). Navy bean yields were increased 36% (Hofstra *et al.*, 1978), white bean by 24% (Temple & Bisessar, 1979), tomato by 30% (Legassicke & Ormrod, 1981), potato by 35% (Bisessar, 1982) and tobacco by 20% (Bisessar and Palmer, 1984).

In the USA, EDU treatment increased yields by 31% in New Jersey (Clark *et al.*, 1983) and 19% in California (Foster *et al.*, 1983). The validity of such work is supported partly in that little or no yield increase is achieved with cultivars that are genetically tolerant of ozone. Nor are there any effects when ozone is not present or has been filtered from the air (Legassicke & Ormrod, 1981; Foster *et al.*, 1983).

Table 9.2. Estimated percentage yield loss based on seasonal average 12-hour
ozone concentration[1]

Crop	3-month, 12-hour average ozone concentrations			Ambient ozone (1984)
	40 ppb	50 ppb	60 ppb	
Lemons	12.7	20.8	22.9	28.3
Dry beans	10.5	16.9	22.7	27.2
Onions	14.2	20.8	22.9	23.2
Grapes	9.4	15.2	19.5	20.8
Cotton	6.6	11.1	15.3	19.6
Oranges	8.9	14.7	18.2	19.3
Rice	6.8	9.2	10.2	10.4
Alfalfa	4.3	7.4	7.5	7.6
Sweet corn	3.8	5.4.	6.1	6.1
Tomatoes	0.6	1.6	2.6	4.5
Silage corn	0.5	1.2	2.2	3.5
Field corn	0.4	1.0	1.5	1.7
Wheat	0.8	1.4	1.7	1.7

[1] Adapted from CARB Staff Report, 1987.

However, these results are not always consistent with those obtained in fumigation studies. When random plots of two soya bean cultivars were treated with EDU over a three-year period, yields did not differ from the untreated group, although ozone concentration exceeded 120 ppb for as many as 72 hours during the growing season, and the seasonal 7-hour daily mean equalled 62 ppb (Smith *et al.*, 1987). The NCLAN models that would have projected yield losses did not reflect the episodic nature of ozone pollution or some of the more important environmental conditions, most notably water stress, that influence plant growth in the field.

MECHANISMS OF OZONE ACTION

Pathways of ozone

Even before the minute flecks denoting ozone injury become visible to the naked eye, physiological and ultrastructural changes are taking place within the cell. Other stresses evoke much the same responses, but those initiated by ozone have some unique characteristics.

Along its course into the plant cell, an air pollutant, in this case ozone, first contacts the cuticle and stomata of the leaf. That portion of the gas contacting the cuticle interacts thereon and may cause erosion or breakdown of that surface. The portion of greatest concern, though, is that passing through the stomatal aperture. Some of the ozone comes into contact with the guard cells surrounding the stoma, causing a decrease in their turgidity, possibly related to the ozone-induced loss of potassium from the cell; thus causing them to close, precluding or reducing further entry (Mansfield, 1973). The underlying response may be the disruption of membrane permeability.

Passing through the stomata, the ozone enters into the substomatal intercellular spaces, where it is highly reactive and dissolves in the water of the moist cellular surfaces. Ozone decomposes rapidly, releasing molecular oxygen together with a number of free radicals and ions (e.g. HO_2^-, HO^+, OH^-, O^- and OH_2^-) which can subsequently oxidize various cellular metabolites and affect a number of membrane constituents such as sulfhydryl (SH) groups, amino acids and unsaturated fatty acids (Heath, 1975). When radicals are not neutralized they are free to attack other substrates, including enzymes, perhaps altering their molecular structure. Some of these may be associated with photosynthetic and other metabolic pathways (Bennett *et al.*, 1984).

Along its pathway through the inner air spaces, ozone may react with olefinic compounds. Much may be scavenged by ascorbic acid before reaching the plasmalemma (Castillo & Greppin, 1986). Ascorbate may react with superoxide radicals, formed from ozone, to produce dehydroascorbate. The amount of ascorbic acid in the cell wall solution may provide a major ozone sink and be responsible for the tolerance of some plants to ozone.

Ozone not reacting with water contacts the cell wall, which is not particularly reactive. Passing through it, ozone can act on the plasmalemma, the first major barrier. Here, reaction is most likely with the polypeptide chains exposed at lipid bilayers of the plasma membrane (Mudd *et al.*, 1984). Amino acids in the proteins of the membrane are also vulnerable, especially the sulfhydryl groups of cysteine and methionine.

Ozone passing through the plasmalemma remains free to react with cytoplasmic components, organelles and their membranes. The double membranes of the endoplasmic reticulum would be most readily attacked, but it is at the chloroplast where the impact is the greatest and apparently precedes injury to the plasmalemma and other membranes. The earliest biochemical reaction site is apparently the membranes of the chloroplast Ledbetter *et al.* (1959). The thylakoid membrane is especially sensitive, probably due to oxidation of the sulfhydryl groups (Thomson *et al.*, 1966). It is the thylakoid membrane in which are embedded the enzymes and cofactors that facilitate the light reactions of photosynthesis; so it is no wonder that any disruption of this structure will seriously affect photosynthesis. It is on the thylakoid disc that we find the complexes that contain the chlorophyll molecules and other pigments and proteins that form the photosystems that gather the light energy. It is important to understand that the *structure* of the photosynthetic apparatus is essential to photosynthesis. Chlorophyll in solution, outside a chloroplast, does not photosynthesize.

Mitochondria and their membranes are subject to attack, but are less responsive than the chloroplasts. Other cytoplasmic substances may be affected, including proteins, organic acids and carbohydrates, but these are less critical and may involve secondary reactions (Koziol, 1984).

Ultrastructure

The earliest identifiable ultrastructural changes reported are to the chloroplasts. The envelope loses its integrity, the chloroplasts become disorganized, and their contents aggregate against the cell wall (Hill *et al.*, 1961). The irregular shape of the ozone-exposed chloroplast is especially notable, together with the increase in granulation and electron density of the stroma (Thomson *et al.*, 1974). Crystalline fibril arrays composed of fibers form during ozone exposure much as from PAN. These are located both near the envelopes and in the vicinity of the grana and stroma thylakoids. Swelling of mitochondria similar to the PAN response has been reported for tobacco leaves fumigated with ozone (Swanson *et al.*, 1973). Electron-dense accumulations are reported in association with the boundary envelopes (Thomson *et al.*, 1974). The effects of ozone on the cytoplasm are more general, involving its aggregation into a dense, collapsed mass around the periphery of the cells (Pell & Weissberger, 1976).

Indentation, or invagination, of the outer chloroplast membrane also has been described together with the crystalline arrays (Thomson *et al.*, 1974). However, this may be a more general response to membrane injury and changes in permeability. Other stresses that cause a similar response include nitrogen dioxide (Thomson, 1975).

David Tingey at the EPA Terrestrial Research Laboratory, and G.E. Taylor (1982), proposed that ozone indirectly induced free radical formation in chloroplasts, making this the site of initial response to ozone. The plasmalemma is subsequently injured together with fine structure changes. At higher ozone concentrations these include the increased electron density in the chloroplast stroma, formation of crystalloids and rupturing of the outer chloroplast membrane. At concentrations of 150 ppb for 8 hours there is a reduction in chloroplast size, disintegration of the thylakoids and a decrease in numbers of ribosomes (Toyama, 1976). Plastoglobules and phytoferritin increase in the chloroplasts. The vacuole ruptures and the organelles are destroyed.

Ultimately other membranes, the tonoplast and endoplasmic reticulum, are destroyed (Pell & Weissberger, 1976). Collectively, the changes result in cellular destruction and death preceding the appearance of general externally visible expression of injury.

Biochemical and physiological effects

Photosynthesis

Photosynthesis is far from the only metabolic process influenced by ozone, but it is intimately linked to productivity and has deservedly received the most attention.

It is easy to say that ozone affects photosynthesis, but many factors enter into photosynthesis. Which of these is most sensitive to ozone? And which is the first to be affected and how might this influence subsequent interactions?

The ozone concentrations first impairing photosynthesis are lower than those at which yield reductions can be measured. Photosynthesis in the sensitive flag leaves of oats, for instance, is suppressed at concentration of 70 ppb (Forberg *et al.*, 1987). White pine proved even more sensitive, and net photosynthesis was reduced 15–20% following 3 months exposure to as low as 40 ppb ozone. No change was reflected in the dry weight of the pine seedlings (Reich & Schoettle, 1987).

Photosynthesis and related processes take place in the chloroplast, and it is here that we must first direct our attention. Ozone affects so many related processes, it is difficult to decide which comes first. One vital process inhibited is electron transport in the water-splitting light reaction, whereby oxygen is released and energy is made available to drive the so-called dark reaction, in which carbon dioxide is reduced (hydrogen is added) and carbohydrates are formed. This is accomplished by the coenzyme nicotinamide adenine dinucleotide phosphatase (NADP), which after capturing an electron from chlorophyll can accept a hydrogen ion from the splitting of water molecules, becoming NADPH or reduced NADP. NADPH then takes part in the sugar-building reactions of the carbon cycle. This inhibition is demonstrated in isolated spinach chloroplasts (Coulson & Heath, 1974). In this process ATP production declined concurrently with both electron transport and the H^+ gradient. This was the first detectable effect of ozone on photosynthesis; then came inhibition of electron transport between the photosystems (Schreiber *et al.*, 1978).

Membrane permeability, especially in the chloroplast, also is altered (Nobel and Wang, 1973). It is concluded (Mudd *et al.*, 1984) that 'the amino acids of the polypeptide chains external to the membrane are susceptible to oxidation by O_3 because of their accessibility and inherent susceptibility to oxidation'. Despite the similar susceptibility of the double bonds of fatty acids to oxidation, they are less readily accessible to ozone and therefore less oxidized. Oxidation of sulfhydryl groups are most likely to account for loss of enzymatic activity, and the accessibility of the sulfhydryl group on the enzyme will influence its likelihood of reacting. Reactions with sulfhydryl groups would reduce the secondary reactions of photosynthesis. A number of photosynthetic, as well as other, enzymes similarly may be adversely affected.

Ribulose bisphosphate (RuBP) is the five-carbon compound that is the short-lived precursor of 3-phosphoglycerate (PGA), which is the first stable intermediate in the pathway towards producing glucose and fructose. Simplistically, RuBP is the electron acceptor for carbon dioxide. Ribulose bisphosphate carboxylase catalyzes the first reaction in this cycle. It is a key enzyme in a pathway central to life and accounts for up to 50% of all the protein in chloroplasts. It may well be nature's single most abundant protein. To what extent is it affected by ozone or the related free radicals?

Ozone has been found to reduce the activity of the carboxylase enzyme that is vital to carbon dioxide fixation and thereby limits production of sugars (Nakamura & Saka, 1978). Inhibition of RuBP carboxylase can begin within 48 hours of exposure to ozone (Pell & Pearson, 1983).

Ozone also affects and can destroy chlorophyll, leading to reduced photosynthesis, but mostly at higher concentrations that involve visible injury (Runeckles & Resh, 1975; Knudson *et al.*, 1977). While net photosynthesis can be impaired without the development of visible symptoms, research first suggested that photosynthesis tends to return to normal when the exposure stops (Pell & Brennan, 1973; Hill & Littlefield, 1969). This is not always the case though, and exposure of ponderosa pine trees to 150 ppb ozone for 30 days reduced net photosynthesis 10% over controls in the absence of visible symptoms (Miller *et al.*, 1969). Reductions in photosynthesis were also demonstrated in citrus (Thompson *et al.*, 1967).

Respiration

Ozone apparently evokes much the same response as other stresses in causing an increase in the rate of respiration (Duggar & Palmer, 1969). One way in which ozone may act is in inhibiting phosphorylation of leaf mitochondria (Lee, 1967). ATP and total adenylate increased immediately following ozone exposure (Pell & Brennan, 1973). It has been postulated that the increased energy comes from lipids and proteins in the cell membranes once the normal carbohydrate reserves are exhausted (Skarby *et al.*, 1987).

Carbohydrate and related metabolism

Following the probable initial effects of ozone on membranes and photosynthesis, a number of secondary responses might be expected. Both increases and decreases in sugars have been reported, depending largely on the ozone concentrations. An increase in soluble sugar content of ponderosa pine needles followed by a decrease in the roots may be especially important.

Changes related to the activity of enzymes in the glycolytic pathyway and stimulation of the pentose phosphate pathway may be responsible (Tingey *et al.*, 1976). This also occurs in diseased and aged plant tissues (Goodman *et al.*, 1967); in other words, it accounts for the premature senescence that often characterizes ozone injury.

Ozone can also affect polyunsaturated fatty acids by oxidative mechanisms; these oxidations in turn can change the properties of membranes (Heath, 1975).

Phenols and related metabolism

Accumulation of isoflavonoids in ozone-treated plants is reported, and it is suggested that this might involve a general stress response as known to be associated with pathogenic infections (Keen & Taylor, 1975). The build-up increased with the appearance of visible symptoms. Enzymes involved in phenol metabolism may increase following ozone exposure together with peroxidase activity (Tingey *et al.*, 1976).

Nitrogen metabolism

Ozone has been found to inhibit both nitrate reductase, and especially nitrite reductase, activity (Tingey *et al.*, 1973). This could influence the nitrogen available for photosynthesis as well as the photosynthetic efficiency. An increase in free amino acids in foliage of several plant species following ozone exposure also has been demonstrated (Craker & Starbuck, 1973). This surplus infers that protein synthesis is impaired. However, effects on protein content are inconsistent, ranging from increases to decreases in various studies (Guderian *et al.*, 1985).

Stress and senescence

Plants subjected to stress produce elevated levels of ethylene (Abeles, 1973). Ozone evokes the same response (Craker, 1971). Ethylene production precedes the appearance of visible symptoms but appears to be associated with many of the phenol and pigment accumulations. Premature senescence associated with ozone is characterized by increases in reactions such as loss of starch (Duggar & Palmer, 1969), proteins (Craker & Starbuck, 1973) and chlorophyll (Runeckles & Resh, 1975), and increases in anthocyanin and polyphenols (Howell, 1974), respiration and ethylene production. Premature leaf abscission and fruit drop also occur (Thompson *et al.*, 1972; Thompson & Taylor, 1969). The effects of ethylene in plants include the promotion of both abscission and fruit ripening.

Assimilate partitioning

The impact of ozone in altering the partitioning of carbohydrates in the plant may be its most significant role (Cooley & Manning, 1987). If a plant is to be healthy, the photosynthetic products must be properly distributed. Ozone generally reduces the amount of dry matter produced, but even if the amount of photosynthate were not reduced the available assimilate is diverted to the leaves and stem at the expense of the root and crown.

The amount of starch accumulating in roots tends to decrease with increasing ozone exposure. Ponderosa pine seedlings exposed to 100 ppb ozone 6 hours per day during the growing season were found to have lower root reserves of soluble sugars and starch in the autumn, which had the potential for restricting initiation of new growth the following spring (Tingey *et al.*, 1976). In one study, however, ozone concentrations above 0.15 μl/l caused significantly greater starch accumulation in pitch pine roots than in the control trees. There were no effects at ozone concentrations of 0.08 or 0.10 μl/l following 8 weeks of exposure to 4-hour daily fumigations.

At ozone concentrations above about 100 ppb, assimilate accumulation becomes greatly depressed and effects on partitioning are not obvious. At lower ozone concentrations of 50– to 100 ppb, storage organs, including roots of trees, are most affected.

The lack of carbohydrates diverted to the roots can be especially critical when mycorrhizae are involved. This is most vital with conifers. The ozone-induced premature senescence of the needles, the primary source of photosynthate, is serious in itself, since less sugar is then available for transport. But the reduced export to the roots and bole may be still more damaging even in the absence of premature senescence.

Less vesicular arbuscular mycorrhizae development occurs if the flow of carbohydrates to the roots is reduced. It is also possible that the reduced carbohydrates could disrupt the normal fungal balance and make the roots more susceptible to pathogens (Manning, 1978).

Ozone apparently affects translocation physiology, i.e. translocation of carbohydrates in the phloem, in ways still unknown, so that more photosynthate is translocated to the young leaves and less to the roots and stem. As flowers develop, more carbohydrates may be translocated to the reproductive organs at the obvious expense to the roots.

SUGGESTED READING

Dempster, J.P. & Manning, W.J. (1988). Response of crops to air pollutants. *Environ. Pollut.* (special issue) **53**, xxiii + 478 pp.

Guderian, R., Tingey, D.T. & Rabe, R. (1985). Effects of photochemical oxidants on plants. pp. 130–346 in *Air Pollution by Photochemical Oxidant* (R. Guderian, ed.). Springer-Verlag, Berlin. 346 pp.

Koziol, M.J., & Whatley, F.R. (1984). *Gaseous Air Pollution and Plant Metabolism.* Butterworths, London, xiii + 466 pp., illus.

Krupa, S.V., & Manning, W.J. (1988). Atmospheric ozone: Formation and effects on vegetation. *Environ. Pollut.*, **50**, 101–138.

US Environmental Protection Agency (1986). *Air Quality Criteria for Ozone and Other Photochemical Oxidants.* Vol. 3, Ch. 6. Env. Criteria Assessment Office. Research Triangle Park, North Carolina. xxvi + 298 pp., illus.

CHAPTER 10

Air Pollutant Interactions

No air pollutant occurs alone. We study the effects of ozone alone, or sulfur dioxide, or fluoride; but air is really a mixture of natural and unnatural gases and suspended particulate and aerosol matter. In a polluted atmosphere one contaminant will most likely be predominant, but others may be present in varying lesser amounts. While most studies are directed towards learning about plant responses to a single pollutant, co-occurrences of pollutants do occur, and the combined presence of two or more pollutants may have different effects on vegetation from those of a single pollutant. Theoretically, the combined effects may be the same as when each pollutant is applied alone (additive), greater than the sum of each applied alone (synergistic) or less than if each were applied alone (antagonistic).

Interaction studies have largely addressed plant responses to a combination of only two pollutants. But combinations of three or more are also possible. Also, variations may occur not only in combinations but in concentrations and exposure regimes. Peak concentrations of different pollutants may not occur at the same time of day. When the possible interactions with other environmental parameters, such as climatic and edaphic factors, are considered, the potential responses are nearly infinite (Ormrod, 1982; Runeckles, 1984).

Several questions must be considered when addressing the collective effects of more than a single pollutant. Many of these concern the mechanism of actions that could be altered. Does one pollutant alter the plant's predisposition to another? Looking at the pollutants *per se*, is the ratio of one pollutant to another important? And how might the relative or sequential doses of pollutant combinations influence plant response?

It is especially critical to recognize that the controlled exposure conditions found in most studies of responses to pollutant mixtures do not necessarily suitably approximate the varied, diverse ambient conditions that occur in the field, both with regard to exposure regimes and other environmental parameters.

122

CO-OCCURRENCES OF AIR POLLUTANTS

Research addressing the combined effects of pollutants would be irrelevant if such co-occurrence were infrequent or lacking. Co-occurrences of ozone and sulfur dioxide are the most common. They occur most notably near coal-fired power-generating plants and other industries where combustion processes generate both sulfur dioxide and oxides of nitrogen. In addition to the sulfur in the coal that is released in combustion, the high temperatures cause atmospheric oxygen and nitrogen to combine to form nitrous oxide (NO), which is then oxidized to nitrogen dioxide (NO_2), which enters into photochemical reactions leading to ozone formation. The air near urban areas also may be high in both ozone and sulfur dioxide, especially where coal still provides a source of heat or power, or where manufacturing processes releasing sulfur dioxide are present (Tingey *et al.*, 1973).

In order to characterize the extent of co-occurrences of different air pollutants, the EPA's air quality data base, SAROAD, the Electric Power Research Institute's Sulfate Regional Experiment (SURE) data, and the Tennessee Valley Authority (TVA) air quality monitoring data were reviewed (Lefohn and Ormrod, 1984).

Co-occurrence is defined as the simultaneous occurrence of two or more pollutants at hourly concentrations averaging greater than 50 ppb. This value was selected by the researchers because lower concentrations were not considered harmful. They identified 135 monitoring sites at which ozone and sulfur dioxide concentrations exceeded 50 ppb. Although most sites had less than ten co-occurrences during a growing season, some had many more. Sites near Los Angeles experienced numerous ozone episodes above 50 ppb, so that whenever sulfur dioxide concentrations rose a co-occurrence was likely. Even here, however, there were only a few hourly sulfur dioxide values above 50 ppb. Co-occurrences appeared in two other areas in the eastern USA, but sulfur dioxide concentrations remained below 100 ppb in one and 200 ppb in the other.

Monitoring stations are located mostly near urban centers; however, since 1978, the numbers of stations designated as remote or rural have increased greatly. Actual forest or mountain sites remain few in number. Yet, where data are available, concentrations in such areas can be quite high (Lefohn & Jones, 1986). In 1983 the National Park Service established a nationwide network to collect hourly mean ozone, sulfur dioxide and NO_x data. Hourly mean ozone concentrations at such sites exceeded 15 ppb over 90% of the time, compared with 50–70% of the time at sites under the influence of local urban sources. When ozone concentrations at the remote sites exceeded 100 ppb, it was generally during the late evening hours after 1900 hours. Sulfur dioxide concentrations at the sites were generally close to the limits of detection, except where point sources were present. No exceedences of the 50 ppb value were detected for sulfur dioxide.

Ambient ozone, sulfur dioxide and nitrogen dioxide data also were obtained at 11 monitoring stations located throughout the Federal Republic of Germany (Lefohn & Mohnen, 1986). Between 1974 and 1983, ozone concentrations exceeded 50 ppb on numerous occasions. Excesses of 100 ppb were noted to have occurred over 300 times at Arzberg, Hof and Selb. Concentrations reached 163 ppb at Waldhof in 1983.

Sulfur dioxide concentrations decreased over the study period but still exceeded 250 ppb at Hof, Selb and Arzberg. The highest hourly sulfur dioxide concentration recorded was 566 ppb at Arzberg in 1984. However, these were mostly in the winter months when ozone concentrations were low.

Thus, we see that meaningful co-occurrences of ozone and sulfur dioxide are infrequent in Europe and the USA alike. They are mostly restricted to areas where a specific sulfur dioxide source is present. It might be inferred then that it is unlikely that synergistic or additive effects of combinations could have broad significance. However, ozone has become so ubiquitous in recent years that co-occurrences with sulfur dioxide are virtually inevitable wherever that pollutant occurs.

EFFECTS OF POLLUTANT COMBINATIONS

Ozone and sulfur dioxide

The potential effects of ozone–sulfur dioxide mixtures were first suspected in the 1950s (Haagen-Smit *et al.*, 1952; Thomas *et al.*, 1952; Middleton *et al.*, 1958), but the interaction between sulfur dioxide and ozone was not clearly demonstrated until 1966, using the sensitive Bel W$_3$ tobacco variety (Menser & Heggestad, 1966). The synergistic action of this combination provided a major impetus for subsequent research on pollutant mixtures. A number of studies with mixtures were conducted in the 1970s at concentrations ranging between 25 and 1700 ppb sulfur dioxide and from 30 to 500 ppb ozone. For the most part the effects of the combinations were additive, but synergism was evident in some instances. Many such studies have been reviewed (e.g. Reinert *et al.*, 1975) that showed synergism occurred largely with tobacco, but the type of response differed greatly among species and depended further on the concentration of each pollutant.

It is important to note that synergism does not occur on tobacco below the threshold dose of ozone alone, but only below that of sulfur dioxide alone (MacDowall & Cole, 1971). It has also been shown that there was an upper concentration limit to the synergistic response (Gardner & Ormrod, 1977).

Extensive studies have been directed towards learning how pollutant mixtures might influence growth and productivity (Reinert *et al.*, 1969;

Tingey & Reinert, 1971, 1973, 1975). Concentrations of 50 ppb of each gas were used in some studies and 100 ppb in others. Exposures were for 6–8 hours per day, 5 days a week throughout the growing season. Crops studied included tobacco, radish, alfalfa, broccoli, cabbage, tomato, onion, brome grass and spinach. Decreases in growth and yield generally were somewhat greater than the additive effects of the single gases, but the differences were not always significant.

More than additive effects were demonstrated for sensitive quaking aspen clones (Karnosky, 1976). Chlorotic mottle of eastern white pine was greatly intensified when sulfur dioxide was also present. Exposures to either gas alone produced 3–4% needle mottling and premature drop, while exposure to the combined gases at the same dose (100 ppb, 8 hours per day, 5 days per week for 4–8 weeks) produced approximately 16% mottling and premature needle drop.

The influence of other environmental parameters must not be ignored. Exposures to ozone and sulfur dioxide can cause significantly more photosynthetic reduction on sugar maple and white ash under high irradiance and humidity than under conditions of low humidity (Carlson, 1979; Dochinger *et al.*, 1970). Presumably, any conditions leading to full stomatal opening predispose plants to greater uptake and subsequent injury.

Plant responses can range from synergistic to antagonistic depending on the concentrations of the gases used, the duration of the exposure and the sensitivity of the species or cultivar (Heagle & Johnston, 1979). Antagonistic effects, where some apparent cross-protection appears, were demonstrated when bean cultivars were exposed to the combined pollutants (Jacobson & Colavito, 1976). Generally, synergism gave way to antagonism as the exposure or dose increased. Thus, it is not surprising to learn that 'intermediate' concentrations of 250 ppb produce additive effects (Reinert & Weber, 1980). Less than additive effects were demonstrated in experiments with white bean and soya bean using concentrations that produced synergistic effects in radish (Beckerson & Hofstra, 1979).

It is important to consider the response of plants exposed to a given concentration of one pollutant when concentrations of a second are increased. Yields of snap beans, for instance, were reduced even at 60 ppb sulfur dioxide in the presence of ambient ozone, although there was no yield loss from ambient ozone alone (Heggestad & Bennett, 1981). Reductions in yields of kidney beans also were greater in the presence of sulfur dioxide and ozone than ozone alone (Oshima, 1978).

What are the specific responses, or symptoms, expressed when plants are exposed to a mixture of gases? This is hard to say when dealing with subliminal responses, but with visible expression we can rely on the classic symptoms. Ozone injury symptoms generally predominate over sulfur dioxide expression (Menser & Heggestad, 1966; Tingey *et al.*, 1973; Heagle *et al.*,

1974; Elkiey & Ormrod, 1979). There are reports, however, where symptoms are distinct from those produced by either gas alone. In one instance, petunia showed undersurface glazing more characteristic of PAN (Lewis & Brennan, 1978); in another, cucumber showed interveinal chlorosis (Beckerson & Hofstra, 1979); and in still another, woody species showed both sulfur dioxide and ozone symptoms (Carlson, 1979).

Subliminal effects are the most critical and, in fumigation studies, have been demonstrated to consist of growth reductions of a magnitude comparable to the combined effects of the individual gases. As with exposures to single pollutants, there is no correlation between the severity of visible symptoms and growth effects (Tingey *et al.*, 1973). Foliar injury seems to be more of an additive response, while at lower concentrations that cause yield losses the response is most frequently synergistic (Tingey & Reinert, 1975).

No published reports could be found of the effects of co-occurrences of sulfur dioxide and ozone under natural conditions in the field. However, on one occasion in Kansas, Treshow (unpublished) observed definitive symptoms of ozone injury on soya beans and other ozone-sensitive species near a sulfur dioxide source, while such symptoms were absent a few miles away from the source. Sulfur dioxide concentrations in one field showing injury had not exceeded 300 ppb, but no ozone data were available.

Explanations for the synergistic and antagonistic responses have been sought in the physiological mechanisms involved. One proposal held that sulfur dioxide in the mixture decreases stomatal resistance, allowing more ozone to enter (Beckerson & Hofstra, 1979). Actually, stomata may be induced either to open or close in response to sulfur dioxide, depending on the species, sulfur dioxide concentration, duration of exposure and environmental conditions at the time of exposure (Black, 1982).

In summary, various studies found that sulfur dioxide decreased stomatal resistance, ozone increased resistance, and sulfur dioxide plus ozone increased resistance much more than ozone alone (Beckerson & Hofstra, 1979). In other words, stomatal function is influenced by too many interactions to explain these plant responses.

Changes in relative humidity, leaf water potential and membrane permeability may be demonstrated among cultivars of differing ozone sensitivity, but they fail to explain differences in sensitivity to the mixtures (Elkiey & Ormrod, 1979). Nutrient relations may have some relevance to synergisms but have received little attention.

Ozone and nitrogen oxides

Despite their frequent co-occurrences as photochemical pollutants, the combined effects of these gases have received scant attention. One reason is the generally low nitrogen dioxide concentrations and the low toxicity to plants of this gas alone. In one study, concentrations of 100 ppb nitrogen dioxide, when combined with 100 ppb ozone, 6 hours per day for 28 consecutive days, reduced height of Virginia and loblolly pine considerably (Kress & Skelly, 1982). Less than additive growth suppression was reported for sweetgum root and total dry weight and white ash root weight.

Other studies conducted with this combination also suggest that the effects are less than additive (Kress, 1980). For instance, exposures of wheat or radish to 100 ppb nitrogen dioxide daily from the hours of 0900 to 1200 sensitized these species to harmful effects of 100 ppb ozone administered from 1200 to 1800 (Runeckles *et al.*, 1978). These conditions are comparable to those typically occurring in urban areas. Nitrogen dioxide alone stimulated top growth, while ozone alone inhibited it. However, the growth-stimulating effects of nitrogen dioxide more than compensated for the ozone impact.

Greater than additive effects were reported using $^{13}CO_2$ (Okano *et al.*, 1984). A mixture of 2.0 ppm nitrogen dioxide and 2000 ppb ozone reduced the $^{13}CO_2$ fixation of leaves and altered the pattern of assimilate distribution more drastically than would be expected by either pollutant alone. The amount of labeled assimilates translocated to the roots and lower stem was reduced by 85% and 80%, respectively. These studies demonstrated the complexity of pollutant interactions and that sequential exposures might be as important as concurrent exposures to pollutants (Runeckles & Palmer, 1987).

Ozone and PAN

Peroxyacetyl nitrate (PAN) and ozone are both photochemical pollutants and would be expected to occur together. Yet few studies have addressed the effects of such co-occurrences, and their results have been inconsistent. In one study (Kress, 1972; Kohut, 1972) exposures to the two gases in early summer caused visible injury on hybrid poplar that was synergistic, while later in the year they become additive and subsequently less than additive. Later studies (Kohut & Davis, 1978) described only synergistic effects on poplar and pinto bean. However, synergistic responses were limited to the upper leaf surface, while the response of the lower surface was antagonistic.

Ozone and hydrogen fluoride

The paucity of studies of this combination reflects the infrequency of their occurrence. Interactions shown by these studies are quite variable and reflect largely a dose dependency (Runeckles, 1984). D.C. McCune (1983) at the Boyce Thompson Institute primarily looked at how fluoride uptake was influenced by ozone. Uptake was reduced in alfalfa and ryegrass but was not affected on tomato, maize, timothy, pinto bean, mint or *Coleus blumei*.

Hydrogen fluoride exposures ranged from 0.9 to 19 ppb for as long as 30 days, and ozone regimes from 50 to 200 ppb for 4–8 hours daily for 9–12 days.

Synergistic injury responses were noted for leaf injury to *Coleus* and mint. Plant biomass of tomato was reduced additively, but hydrogen fluoride reduced the amount of ozone-induced growth reduction. Intercostal necrosis, not characteristic of either pollutant alone, occurred on tomato.

Nitrogen oxides and sulfur oxides

Among combinations not involving ozone directly, sulfur dioxide and NO_x are the most likely to co-occur, especially in the vicinity of coal-fired power-generating plants (Ashenden, 1979b). Where oxides of nitrogen occur, however, ozone also is likely to be present so, realistically, effects of the three gases together should be studied.

Visible injury was produced in six species following a single 4-hour exposure to mixtures of nitrogen dioxide and sulfur dioxide at concentrations of less than 250 ppb (Tingey *et al.*, 1971). No visible injury was produced by either pollutant alone at a much higher dose.

Synergistic injury responses, causing symptoms different from either pollutant alone, were produced on geranium, petunia and tomato at 300 ppb sulfur dioxide and 500 ppm nitrogen dioxide (DeCormis & Luttringer, 1977). Much lower concentrations of 60–80 ppb nitrogen dioxide and sulfur dioxide caused significant reductions in leaf area and plant dry weight of the grasses *Dactylis glomerata* L. and *Poa pratensis* L. (Ashenden, 1979a).

Soya bean leaf, stem and root dry weights were decreased additively following exposures to nitrogen dioxide and sulfur dioxide. Reductions up to 32% were reported at 200 ppb nitrogen dioxide and 300 ppb sulfur dioxide exposures for 3 hours every other day for 30 days (Klarer, 1982).

Synergistic yield reductions were found for soya beans exposed to 60–400 ppb nitrogen dioxide and 130–420 ppb sulfur dioxide. Exposures to nitrogen dioxide alone had no effect, and exposures to the sulfur dioxide decreased yields by 6%. The combination resulted in decreases of 9–25% in the two years of study (Irving *et al.*, 1982). Ozone present in the ambient atmospheres used in the study ranged from 6 to 95 ppb and may have had some interactions. Similar studies (Amundson and Weinstein, 1980, 1981) showed that sulfur dioxide concentrations of 100 and 300 ppb and nitrogen dioxide at 100 ppb, 4 hours daily for 14 days during the pod-fill stage, caused both yield reductions and early senescence. The gases alone produced no effect.

Of the woody plants, growth of poplars was reduced 37% by continuous 8-week exposures to 60 ppb sulfur dioxide and 60 ppb nitrogen dioxide. No reductions were caused by either pollutant alone (Whitmore *et al.*, 1982).

These and other studies all suggest that sulfur dioxide and nitrogen dioxide are frequently dependent on one another to suppress growth. The mode of action for this interaction remains highly conjectural. Nitrogen oxides, however, are known to reduce activity of nitrite reductase (NiR), while sulfur dioxide has little effect. Alan Wellburn (1982) has proposed that the presence of sulfur dioxide prevents induction of additional NiR by nitrogen dioxide, which would normally lead to ammonia and amino acid synthesis.

Membrane damage from the inability of nitrogen dioxide to detoxify itself in the presence of sulfur dioxide is also well documented. As sulfur dioxide increases, sulfite accumulates (Malhotra & Hocking, 1976). This, plus nitrite accumulation, would be especially disruptive to the integrity of membranes. Such a disruption could allow ozone and other pollutants to enter the cells more freely to foster synergistic responses.

Nitrogen dioxide, ozone and sulfur dioxide

Studying combined effects of three pollutants increases the complexity appreciably, both in experimental design and possible combinations of exposure regimes. Realistic combinations of nitrogen dioxide, ozone and sulfur dioxide generally tend to suggest a synergistic action. Turf grasses exposed to these gases at a concentration of 150 ppb each, for instance, showed more leaf injury and greater reduction in leaf area compared with exposures to the pollutants applied singly (Elkiey & Ormrod, 1980). On the other hand, studies with sweet corn and rice suggested antagonism by sulfur dioxide (Yamazoe & Mayumi, 1977).

Mixtures of the gases at concentrations below the national ambient air quality standard for the gases (e.g. 120 ppb for ozone) caused substantial growth suppression of loblolly pine and American sycamore (Kress *et al.*, 1982, 1982b).

Nitrogen dioxide alone at ambient concentrations generally does not affect plants. However, when sulfur dioxide and/or ozone are present, phytotoxic effects have been shown. The effects are mostly additive or somewhat greater. Suppressed growth and yield are in the 5–20% range even at concentrations below the ambient air quality standard.

Sulfur dioxide and fluoride

These pollutants are emitted by a number of industrial processes. Such emissions have been subject to considerable regulation and control in many countries in recent years, which has reduced the probability of co-occurrences.

In at least one study, no influence of one gas on the other was found in growth and development of citrus (Matsushima & Brewer, 1972).

Emphasis on hydrogen fluoride–sulfur dioxide interaction generally has been directed towards fluoride accumulation. For example, the significant decrease in fluoride accumulation under a sulfur dioxide regime has been noted (McCune, 1983). Studies with ryegrass, gladiolus and sweet corn showed that the effect was dependent on the concentrations of sulfur dioxide and hydrogen fluoride, occurring only when sulfur dioxide concentrations exceeded 500 ppb in the presence of a very high 12.5 ppb hydrogen fluoride, or 400 ppb sulfur dioxide at 5.5 ppb hydrogen fluoride. Effects on fluoride accumulation were suspected to be explained by differences in sulfur dioxide-induced stomatal closure.

Studies summarized in 1983 (McCune, 1983) generally reflected additive effects with each pollutant acting independently. Synergistic effects of increased lesions were noted for barley or maize, but only following exposures to 80 ppb sulfur dioxide and 0.8 ppb hydrogen fluoride for 27 days.

An interesting study in Switzerland (Keller, 1980) suggested a synergistic reduction of carbon dioxide uptake of Norway spruce by 75 ppb sulfur dioxide in the presence of increased tissue fluoride concentrations. Sulfur dioxide also increased fluoride accumulation in roots and needles.

SUGGESTED READING

Kress, L., Skelly, J.M., & Hinkelman, K.H. (1982). Growth impact of O_3, NO_2 and/or SO_2 on *Pinus taeda*. *Environ. Monit. Assoc.*, **1**, 229–239.

Runeckles, V.C. (1984). Impact of air pollutant combinations on plants. pp. 239–258 in *Air Pollution and Plant Life* (M. Treshow, ed.), Wiley, Chichester. xii + 486 pp., illus.

US EPA Criteria Document (1986). *Air Quality for Ozone and Other Photochemical Oxidants*. Vol. 3, Ch. 6. Environ. Criteria Assess. Office, RTP, North Carolina. 298 pp.

CHAPTER 11

Environmental Interactions

It is axiomatic in ecology that everything affects everything else. Nowhere is this more true than with how environmental parameters influence the impact of air pollutants and with the sensitivity of the plant. Abiotic factors, notably water status, temperatures, light regimes and mineral nutrition, are most critical, but biotic factors, including fungi, bacteria and viruses, also show interactions with pollutants. Even short-term variation in these agents, if they are synchronous with a pollutant episode, can influence the sensitivity of plants to a pollutant (Ting & Duggar, 1971).

ABIOTIC STRESS

Moisture relations

Relative humidity

The water status of the plant is particularly important and is influenced both by soil and atmospheric moisture. The atmospheric moisture, or relative humidity, has a most obvious effect through its influence on leaf turgor, thereby controlling stomatal opening and gas exchange (Tingey *et al.*, 1982; Guderian, 1977; Unsworth *et al.*, 1976). Interactions with ozone have been studied most intensively. When plants are dry, the stomata close, and less ozone can enter the leaf; when the relative humidity is high, stomata are more open and the ozone exposure is increased. Increasing relative humidity from 35% to 73% can increase ozone uptake fourfold (McLaughlin & Taylor, 1981). However, if the relative humidity exceeds 90%, sensitivity to ozone may decrease (Dunning & Heck, 1977).

In another study, daily leaf diffusive resistance (LDR) was unaffected at 40% relative humidity, but changed significantly at 80% over the 5 days of the experiment (Jensen & Roberts, 1986). The higher humidity, facilitating

131

stomatal opening, allowed entry of more ozone and subsequent injury to the guard cells. Damage to epidermal cells adjacent to the guard cells, leading to still wider opening of the stomata, caused the LDR to drop. Presence of the pollutant may indirectly affect cell permeability to potassium, which is important in regulating changes in turgor.

The positive relation between relative humidity and ozone injury was established first for tobacco in 1963 (Menser), but the relation subsequently has been documented many times on other crops (Otto & Daines, 1969; Rich & Turner, 1972). Correlation also was established between relative humidity and ozone-induced needle necrosis that developed during needle elongation (Davis, 1970). This also is true with American ash (Wilhour, 1970) and pinto bean but only when plants are grown at a high light intensity (Dunning & Heck, 1977).

Actual uptake of ozone has been shown to be enhanced at a higher relative humidity together with the greater stomatal apertures (McLaughlin & Taylor, 1981).

The implication of this relationship is relevant to the presence of low humidities in arid regions that might partially mitigate ozone effects, but especially in more humid regions where ozone injury would be enhanced. Thus ozone concentrations of 100–150 ppb might be highly injurious in one area, while of little consequence in areas of low humidity.

Soil moisture

Moisture available in the soil would have a similar effect through influencing the stomatal aperture, especially in the short term. This was demonstrated as early as 1952 (Hull & Went, 1952) when fully turgid plants exposed to ozonated hexene were more severely injured than water-deficient plants. Later, it was found that stomata of water-stressed tobacco plants were generally closed, while stomata of well-watered plants remained open even under exposure (Rich & Turner, 1972). This has led to the premise that some protection from ozone could be obtained by subjecting plants to partial water stress. Indeed, such protection could be obtained within a few days, depending on the severity of the stress (Tingey *et al.*, 1982). This has limited practicality to agriculture because water stress will severely limit productivity. Tobacco farmers have been informed over the radio of high ozone episodes so that they could limit irrigation where possible (Dean & Davis, 1972).

In California, following air pollution episodes, extensive injury has been observed on well-watered plants, in contrast to a general absence of injury on non-irrigated plants (Taylor, 1974; Temple *et al.*, 1985). At the cellular level water affects the course of reactions and the basic responses (Mudd, 1975).

Ozone-drought stress interactions were studied in the field for cotton, barley, alfalfa and lettuce (Temple *et al.*, 1985, 1986). Drought stress was imposed by manipulating the amounts of irrigation water applied. Plants were exposed to naturally high-ozone ambient air and filtered air and to a gradient of ozone concentrations. Cotton yields were significantly reduced by reductions in soil water, but there was no loss in yield from ozone under the stress conditions. In a more humid season, in the absence of moisture stress, ozone caused significant yield suppression (King, 1987).

Similar studies with alfalfa showed that in some cuttings both ozone and soil moisture deficit significantly reduced alfalfa yields. Results from June and July of one year, when evaporative demand was greatest, showed that there was no yield loss in response to ozone when plants were drought stressed (Temple *et al.*, 1988).

Under natural agricultural conditions it would seem that, although moisture stress can mitigate some of the adverse effects of ozone, the stress itself would be more damaging.

Mineral composition

Plant nutrients

The complex interaction and influence of nutrition on plant response to air pollution have led to conflicting results over the years. Perhaps some of the conflicts are due to the influence of such variable soil factors as pH, water relations, soil type or availability of nutrients that might influence availability of others.

Sulfur dioxide responses to nutrition have received the earliest and most attention. In one study (Bleasdale, 1952) perennial ryegrass was exposed to ambient atmospheres in which sulfur dioxide concentrations ranged between 20 and 252 $\mu g/m^3$ and up to 504 $\mu g/m^3$ (approx 200 ppb) in another. Plants were fertilized with low or high levels of nitrogen, phosphorus and potassium. In each experiment sulfur dioxide impaired growth least in plants having the higher fertilization levels. No visible leaf injury from sulfur dioxide developed.

Just the opposite was found for tobacco and tomato (Leone & Brennan, 1972). Sensitivity, as determined by lesions, was greatest in plants having ample nitrogen and decreased with either excess or deficient amounts of fertilizer. Decreased sensitivity appeared to be related to increased stomatal resistance since deficiencies are known to increase such resistance. However, the direct effect of sulfur dioxide in influencing stomatal resistance, depending on concentration, complicates the sulfur dioxide–nutrient interactions and must not be overlooked.

A review of much of the conflicting data suggests that many plants may be more sensitive to sulfur dioxide under conditions that give a low rather than high rate of growth (Cowling & Koziol, 1982).

Fluoride–nutrient interactions were studied extensively in the 1960s, especially with regard to calcium. Here, the action was unique in that fluoride tended to bind the calcium in developing fruit, most notably peach. This caused a calcium deficiency, which was expressed as a premature softening of the suture area. The disease was known as soft suture or suture red spot, and could be controlled by applications of lime (Benson, 1959). The binding of fluoride with calcium and magnesium also explains why leaf injury is most severe when these elements are deficient (MacLean *et al.*, 1969).

In more traditional interactions, tomato plants became increasingly more sensitive to hydrogen fluoride as the phosphorus supply increased (Brennan *et al.*, 1950; Daines *et al.*, 1952). Sensitivity also increased with higher nitrogen levels (Adams & Sulzback, 1961; McCune *et al.*, 1966).

Another consideration, in the case of fluoride, is the influence that nutrition has on fluoride uptake and accumulation. Fluoride accumulated almost linearly with increases in either phosphorus or potassium, despite little to no effect on injury (Guderian, 1977). Nitrogen had the reverse effect where fluoride accumulation and injury were greater, in spinach grown with low nitrogen. From the view of reducing fluoride content in forage, optional nutrition would stimulate growth and thereby dilute the fluoride content in the more rapid, lush growth (Treshow, 1970).

Early studies demonstrated that oat, barley, spinach and lettuce were more sensitive to ozonated hexenes when nitrogen levels were optimal (Middleton, 1956). Addition of phosphorus alone decreased ozone injury, but addition of potassium alone usually increased oxidant injury in the presence of low but not high nitrogen. Addition of both phosphorus and potassium significantly decreased injury to spinach. On the other hand, potato leaves deficient in nitrogen were most sensitive to ozone injury. Grapevines also were most sensitive at low nitrogen concentrations (Kender & Shaulis, 1976).

Ozone-sensitive tobacco provides a useful indicator of nutrient–ozone interactions, and the characteristic flecking has been found in the field to be greatest at low nitrogen levels and to decline as the supply was increased (Menser & Street, 1962; Heagle, 1979). In greenhouse sand culture studies, however (Menser & Street, 1962), the reverse was true, with the least injury in nitrogen-deficient plants. Nitrogen content may act on ozone sensitivity through its influence on sucrose content, in that if nitrogen-deficient plants are high in sucrose, ozone sensitivity would be expected to decrease (Lee, 1965).

Leaf injury of tomato plants exposed to 320–960 μg/m^3 (163–489 ppb) ozone for 3 hours decreased when phosphorus was deficient, but there was little effect on radish (Leone & Brennan, 1970). In a later study, ozone injury on tomato foliage was reduced by potassium deficiency (Leone, 1976). This may involve the effect of potassium on the stomatal mechanism. Elevated sulfur nutrition seems to have the opposite effect in providing some protection against ozone toxicity (Adelpipe *et al.*, 1972).

Salinity

A high salt concentration in the soil solution provides a stress to most agricultural plants that can produce a physiological drought. This may reduce the water uptake by the roots and increase stomatal resistance, reducing pollutant uptake. Salinity also may act directly by altering metabolic processes, including photosynthesis. Hence the interactions can be quite complex (Levitt, 1980).

Increases in salinity alone cause reduced growth; as ozone concentration is increased, the effect of salt on growth diminishes, and there is less growth reduction from ozone (Ogata & Maas, 1973; Hoffman *et al.*, 1975). Another example of this occurred when 4000 ppm sodium chloride was added to young red maple trees. Growth was impaired, but the addition of 250 ppb ozone (491 $\mu g/m^3$) had no further impact (Dochinger & Townsend, 1979).

Trace elements

Interactions of pollutants with trace element in plants have been reviewed (Ormrod, 1984; Smith, 1984). Ozone toxicity is enhanced by levels of cadmium or nickel that are not deleterious in themselves (Ormrod, 1977). On the other hand, when growth is suppressed by the metals, ozone injury is diminished. Ozone sensitivity of garden cress, lettuce and pinto bean is enhanced by increased zinc concentrations (Ormrod, 1984).

Interactions of trace elements with sulfur dioxide also have been demonstrated (Krause & Kaiser, 1977). Foliar injury caused by trace elements, for instance, is more severe when sulfur dioxide is present. This could be related to the increase in sulfur content or the sulfur dioxide-induced changes in membrane permeability that result in the release of bound, toxic cations that could enter the cytoplasm.

Cadmium interacts with sulfur dioxide in affecting photosynthesis and transpiration of silver maple leaves (Lamoreaux & Chaney, 1978). The cadmium-induced suppression of the processes was reduced more when sulfur dioxide was present. Stomata resistance also was increased by copper, thereby decreasing sulfur dioxide uptake and perhaps providing plants in mining and smelting areas with some tolerance to sulfur dioxide (Ormrod, 1984).

Temperature relations

Temperatures strongly influence responses to air pollutants. Early work indicated that temperatures favorable to good plant growth also tended to be most conducive to pollutant injury (Taylor *et al.*, 1960; Heck *et al.*, 1965). As the metabolic rate increases with temperature, so does sensitivity to ozone. The optimal temperatures for growth of a given species may play an important role. A species favored by cool weather, such as radish, is most sensitive to ozone under cool conditions (US Environmental Protection Agency, 1986). Pinto bean plants, a high-temperature species, sustain more foliar injury at 32 °C than at 24 °C (Miller & Davis, 1981). However, tobacco expressed less injury at 32 °C compared with 16 °C, 21 °C or 27 °C (Dunning & Heck, 1977).

Other studies showed that plant sensitivity to ozone increased with temperatures over a range of 3–30 °C (Dunning *et al.*, 1974; Dunning & Heck, 1977; Davis & Wood, 1973), but sensitivity to ozone decreased above 30 °C (MacDowall, 1965). High temperatures sensitized the plants earlier, but the effect soon diminished as the younger leaves developed (Juhren *et al.*, 1957). Responses have been found to vary among cultivars (Adelpipe & Ormrod, 1974). Low pre-exposure temperature enhanced ozone-induced growth suppression in one cultivar, while low post-exposure temperature enhanced suppression in another. Plant sensitivity can be increased by high night and low day temperatures (MacDowall, 1965).

Despite a tendency for greatest sensitivity at optimal growth temperatures, the response also depends on the ozone concentration (Dunning *et al.*, 1974). While 1-hour exposures over a temperature range of 16–27 °C had no significant effect on the amount of injury on pinto bean or tobacco, injury at 32 °C was reduced on tobacco but enhanced on pinto bean.

Under field conditions corn was relatively undamaged by ozone when temperatures were less than 32 °C, but significantly injured when temperatures rose above 32 °C (Cameron & Taylor, 1973).

Far more important is the effect air pollutants have on the impact of low-temperature injury (Huttunen, 1978; Munch, 1933; Havas, 1971). The impact of sulfur dioxide in reducing resistance to winter injury was first described by Munch (1933), and subsequently by several others, as reviewed by Huttunen (1984) at the University of Oulu. The problem is most critical with conifers, and even very low sulfur dioxide concentrations during the winter months serve to predispose the trees to winter injury. Ozone may have a similar effect, especially by delaying dormancy and hardening.

Winter crop plants appear to be more sensitive to sulfur dioxide in winter than in summer (OECD, 1978), and resistance to water deficit and winter drought has been noted (Halbwachs, 1968; Huttunen, 1984). Apparently, plant cells injured by even low pollutant concentrations cannot respond adequately to environmental stresses, and low temperatures can provide the final killing agent (Huttunen, 1984). Normal hardening by plants to adjust to winter conditions may be seriously impaired when stressed by pollutants, even when visible injury is not apparent. The total activity of peroxidase may be most significant. Both qualitative and quantitative changes in peroxidase during hardening have been correlated with air pollution conditions. The reduced glucose and sucrose content of spruce needles exposed to sulfur dioxide in the fall may have a role in upsetting the timing of the hardening process (Materna, 1972). Erosion of leaf surface and stomatal wax by sulfur dioxide and ozone also are considered to be critical (Huttunen & Laine, 1981).

Light regimes

Given the fundamental importance of light to plants, it is not surprising that light is also critical to the impact of air pollutants on plants. In many cases harmful effects occur only when light is present. The most obvious interaction is the general closure of stomata in the dark. A 3-hour dark treatment prior to exposure of tomatoes to ozone resulted in decreased injury of foliage (Adelpipe *et al.*, 1973). Light conditions after exposure also can be important. PAN injury requires light before, during and after exposure. A dark period as brief as 15 minutes before a PAN exposure in light can reduce injury. In one study, injury from PAN occurred only when a 2–4-hour exposure to light followed the pollutant exposure (Taylor *et al.*, 1961). Ozone is more injurious when light intensity is high (Juhren *et al.*, 1957; Dunning & Heck, 1977). On the other hand, leaves are more sensitive to nitrogen oxide injury under low light or dark conditions (Taylor, 1973).

With ozone there is no absolute light requirement for injury to develop. Light before, during or after exposure makes no difference, but a 24-hour dark period before exposure can reduce ozone sensitivity in cotton plants (Ting & Duggar, 1968). A 24-hour light period can also reduce sensitivity in the case of young Virginia pine (Davis, 1970).

Photoperiod, i.e. the duration of daylight and darkness, also may be important. Annual bluegrass was more sensitive when grown under an 8-hour than a 16-hour photoperiod (Juhren *et al.*, 1957). Other cultivated plants have been found to be more sensitive under an 8-hour day than either a 12- or 16-hour light period (MacDowall, 1965).

It has been proposed that the amount of ozone injury may depend on the wavelength of light (Duggar *et al.*, 1963). Injury was most intense between 420 and 480 nm. It has also been noted that greater amounts of sulfhydryl compounds are synthesized with increasing light intensity (Duggar & Ting, 1968). These compounds are among the most sensitive reaction sites of photochemical oxidants.

Increased plant injury accompanies increased light intensity during ozone exposure (Juhren *et al.*, 1957; Heck *et al.*, 1965). As is discouragingly inevitable though, other studies have shown the reverse (Dunning & Heck, 1977). Plants exposed to injurious concentrations of sulfur dioxide and ozone in the light (16 klx) were severely injured; injury intensity decreased with decreasing light intensity. Injury in the dark was negligible. Low light intensity during plant growth may increase the sensitivity to photochemical oxidants, and higher light intensity during exposure increases injury (Guderian *et al.*, 1985). This effect is less pronounced in the field because high light intensities there are the norm, and high temperatures and water stress may be superimposed to partially close the stomata.

INTERACTIONS WITH BIOTIC STRESS

The biotic factors inherent in the plant itself clearly impose the greatest influence on sensitivity to any pollutant. No other biotic stress comes close to overriding this basic tenet that the genetics of the plant come first. The phenology of the plant, i.e. its stage of development or maturity, also influences its response to air pollutants, although to a lesser degree. But there are biotic factors that impose stress, or more critically disease, and these may be influenced by air pollutants, just as the pollutants may directly influence the pathogen and the impact of the disease (Saunders, 1975).

The interactions of pollutants and plant disease have been previously reviewed comprehensively (Treshow, 1975, 1980; Huttunen, 1984; Heagle, 1973; Manning, 1975). Consequently, we shall summarize only the most pertinent points and general principles.

Photochemical air pollutants

The growth-inhibiting capacity of ozone against microorganisms was observed long before ozone was recognized to be an air pollutant (Hartman, 1924). Thus, once ozone became a pollutant of note, it was natural to consider its possible interaction with plant diseases. Effects on the causal organisms were explored first (Rich & Tomlinson, 1968; Treshow, 1965; Treshow *et al.*, 1969). Sporulating fungus cultures were exposed to ozone at 100–600 ppb (about 200–1200 $\mu g/m^3$) for periods of 0.5–4 hours. Growth characteristics, including pigmentation, were greatly modified (Kormelink, 1967) and varied considerably among the fungi studied. Ozonated cultures had 28% lower lipid content than controls, possibly owing to the destruction of sulfhydryl groups vital to lipid synthesis (Rich & Tomlinson, 1968). Spore production was impaired by ozone, although their viability was not affected. But how would this translate to actual disease interactions in the field?

The most important implication would be to fungi inhabiting largely the surfaces of fruits or leaves. Ozone has been found to inhibit such fruit-rot fungi as *Monilinia* and *Botrytis*, but only at concentrations above 500 ppb (982 μg/m^3) (Ridley & Sims, 1967). Although ozone inhibited *Monilinia* and *Rhizopus* growth on a number of different fruits, the fruit rot was not significantly reduced during storage (Spalding, 1968). The same was true with *Botrytis cinerea* and *Alternaria* rot.

Surface fungi also would include powdery mildew fungi. Ozone has been found to suppress the development of infection of barley and reduce the establishment of a functional host–parasite relationship (Scheutte, 1971). Formation of mature appressoria was delayed but not suppressed following an 8-hour exposure to 1000 ppb ozone. Lower concentrations impaired development of secondary hyphae essential to infection. Others found initially small colonies on ozone-exposed plants but, over time, colonies became larger on these plants (Heagle & Strickland, 1972).

Ozone inhibited hyphal growth and formation of spores of wheat stem rust fungus as well as infection, so might be expected to reduce the inoculation potential of the fungus (Heagle & Key, 1973). However, in none of these instances was the interaction sufficiently great to expect any real control of the disease under ambient field conditions.

A reverse effect was described earlier by University of California scientists, who found that bean or sunflower leaves infected with rust (*Uromyces phaseoli*) were less injured by smog (Yarwood & Middleton, 1954). The same response was found later with wheat leaves inoculated with *Puccinia graminis* f. sp. *tritici*, which showed significantly less ozone injury than non-inoculated leaves (Heagle & Key, 1973). Similarly, a zone of protection from ozone injury surrounding lilac powdery mildew lesions was found (C.R. Hibben, personal communication). On the other hand, infection of soya bean by *Fusarium oxysporum* increased their sensitivity to ozone (Danicone *et al.*, 1987).

Sometimes the impact of ozone is greatest on the host, and the injury provides an infection court for the fungi, as with *Botrytis* on potato (Manning *et al.*, 1969). However, should the fungus be more sensitive than the host to ozone, sporulation and germination can be inhibited (Krause & Weidensaul, 1978). Generally though, plants weakened by ozone are predisposed to weak pathogens. Such was considered the case when needle cast fungi (*Lophodermium pinastri* and *Pullularia pullulans*) were most frequently associated with eastern white pine exposed to ozone (Costonis & Sinclair, 1972).

Pollutants may enhance or mitigate an interaction depending on the ozone concentration and the plant sensitivity. In one study addressing interaction, exposing ponderosa and Jeffrey pine seedlings to ozone enhanced infection by the wood-rotting fungus *Fomes annosus* (Miller & Elderman, 1977).

Ozone exposure also increased the susceptibility of pine stumps to colonization by *Fomes*. The increased decay susceptibility was borne out further in laboratory studies.

Since ozone has a pronounced effect on cellular metabolism, it would not be surprising to learn that responses to virus diseases might also be affected. An example of this is that during the winter months typical ozone toxicity symptoms consistently develop on tobacco plants free from tobacco mosaic virus (TMV), but never on those infected with the virus (Brennan & Leone, 1969). However, during the summer, both virus-free and virus-infected plants are equally damaged by ozone. It was learned earlier that virus-infected plants had a higher nitrogen content, which may have been related to the ozone resistance (Leone *et al.*, 1966).

Sulfur dioxide

The fungicidal properties of sulfur have been known and utilized for many decades to control fungi. But this involves concentrations hundreds of times greater than where sulfur dioxide occurs as an air pollutant (Treshow, 1975).

The situation in the field may not be the same. To learn if it was, several hundred trees were examined in and outside of an area near a smelter where pollution injury, largely from sulfur dioxide, had been severe (Scheffer & Hedgecock, 1955). There was a clear difference in the abundance of certain fungi. Rust fungi, *Melampsorella cerastii* and *Peridermium coloradense*, and species of *Phragmidium*, *Melampsora* and *Gymnosporangium*, were almost absent near the smelter, although abundant in surrounding areas.

It would seem logical that trees weakened by sulfur dioxide would be more vulnerable to weaker pathogens that thrive on such trees. Near one Montana smelter, Douglas fir and pine trees were infected with the wood-rotting fungi *Polyporus schweinitzii* and *Fomes pini*, but the prevalence and severity were no greater than outside the area (Schaeffer & Hedgcock, 1955). The root and crown fungus *Armillaria mellea*, however, was most prevalent inside the area of sulfur dioxide injury and on weakened trees. Other wood-rotting fungi, including *Glocophyllum abietinum*, *Trametes serialis* and *T. heteromorpha*, also have been found to be most severe on sulfur dioxide-damaged conifers while absent in nearby areas. Species such as *Poria*, *Mycena* and *Polyporus versicolor* occurred only in areas of slight sulfur dioxide damage (Jancarik, 1961).

In the vicinity of a major smelter in Ontario, Canada, fewer white pine trees showed heart rot or symptoms of blister rust (*Cronartium ribicola*) than at more remote sites (Linzon, 1967).

Studies conducted at increasing distances from the industrial town of Torun in Poland, having one major sulfur dioxide source, dealt with economically significant diseases of a Scotch pine forest (Grzywacz & Wazny,

1973). Results showed that *Armillaria* was present in 3.7% of the trees in areas damaged by sulfur dioxide, compared with 1.4% in all forests. The needle cast fungus *Lophodermium* also tended to be more damaging in the industrial areas. Decomposition fungi were less abundant where sulfur dioxide concentrations were highest.

Abundance or absence of fungi near pollution sources is inconsistent even within groups, which might suggest inadequacies in sampling techniques. Populations of one fungus causing needle cast (*Lophodermium pinastri*) were diminished, while populations of another (*L. piceae*) on spruce needles injured by sulfur dioxide had a higher incidence near the source (Jancarik, 1961). Higher incidences of the weakly pathogenic *Rhizosphaera kalkhoffii*, causing needle blight of red pine, occurred in an industrial area of Japan (Chiba & Tanaka, 1968). This is more consistent with the expected in that trees weakened by sulfur dioxide, as with other stresses, should be predisposed to infection.

Fluoride

Fluoride may act on host–parasite interaction more clearly than other pollutants in that it can accumulate in foliage to many times the normal concentration. This affects the metabolic processes of the plant and well may do the same to a parasite. This was shown to be the case where sodium fluoride was incorporated into agar in which representative pathogenic fungi were grown (Treshow, 1965). Even concentrations as low as 5×10^{-4} M impaired growth of *Pythium debaryanum*, while concentrations of 1×10^{-2} M suppressed *Verticillium* and *Helminthosporium*. This would be roughly equivalent to foliar concentrations of 22–420 ppm (calculated as leaf fresh weight of sodium fluoride). Lowest concentrations stimulated fungus growth, suggesting that fluoride might further aggravate the intensity of diseases.

Despite such possibly theoretically fungus-controlling effects of fluoride, the altered plant metabolism caused by fluoride could have an even stronger effect in weakening and predisposing the plant to infection. Fluoride effects on virus diseases have been studied to explore this possibility. In one study (Dean & Treshow, 1965) the numbers of local lesions produced by TMV were compared between plants low and high in fluoride. The numbers of lesions generally increased in leaves having up to 300 ppm fluoride. Lesion numbers then decreased as fluoride content increased above 300 ppm.

Studies of fluoride interactions with bacterial and fungal pathogens are even fewer. Tender green beans were exposed to fluoride and filtered air (McCune *et al.*, 1973). Control plants became severely infected with powdery mildew, compared with only a mild infection occuring on plants exposed to hydrogen fluoride at 7 or 10 $\mu g/m^3$ (about 9–12 ppb). These plants contained 399 ppm fluoride.

Fluoride also has reduced the number of uredia per leaf of plants inoculated with bean rust and reduced the severity of the disease on plants exposed to hydrogen fluoride before inoculation (McCune *et al.*, 1973). Birch leaf rust, caused by *Melampsoridium betulinum*, was absent around an aluminum plant in Norway although widespread a few miles away (Barkman *et al.*, 1969).

Summarizing the potential affects abiotic and biotic agents may have in influencing air pollutant effects, we see that, although numerous changes can be demonstrated under laboratory conditions, effects in the field are limited. While they have been noted in a few instances of severe pollution, even these effects have not notably altered basic air pollutant effects.

SUGGESTED READING

Huttunen, S. (1984). Interactions of disease and other stress factors with atmospheric pollution. pp. 321–356 in *Air Pollution and Plant Life*. Wiley, Chichester. xii + 486 pp., illus.

Treshow, M. (1980). Interactions of air pollutants and plant disease. pp. 103–109 in *Effects of Air Pollutant on Mediterranean and Temperate Forest Ecosystems* (J.B. Mudd & T.T. Kozlowski, eds). Academic Press, London. xii + 383 pp., illus.

CHAPTER 12

Forest Decline and Acid Rain

WHAT IS MEANT BY DECLINE?

The term speaks for itself! First, chlorosis of older leaves, often leading to premature defoliation; then younger needles of conifers progressively becoming yellowed and dropping; and finally shoot tips dying back to produce a thinning canopy. Trees are progressively weakened, the rate of growth is reduced, and often, over a period of years, the tree dies, generally beginning at the top. Mature trees are affected most often. This is not a specific picture, and for a good reason. The more detailed symptoms vary in different regions of Europe and North America. It is for this reason that many scientists believe that there is more than one cause of the recent, 'novel' decline, or 'Neuartige Waldschaden' (Ashmore *et al.*, 1985; Wellburn, 1988).

A number of terms are used almost synonymously with decline. The term *dieback*, for instance, involves the death of outer twigs and limbs. It can be a symptom of decline but is not the whole picture. *Disease* is a broader term and involves any departure from normal health. Diseases are caused by any pathogen, although a biotic agent is often implied. *Injury* refers to a local wound or damage, as might be caused by insects or such abiotic agents as frost, lightning or hurricanes (Treshow, 1970).

We should note that forests have always 'declined'. Pathogenic fungi have been responsible in many instances, insects in other cases, air pollutants in some cases and climatic stresses in still others. None of the classic, familiar explanations seemed valid in the 1980s, at least not any one alone. The new, or recent, decline has many unique features, such as appearing suddenly and showing symptoms on a number of different species, both deciduous as well as coniferous, and occurring over broad geographical areas, and involving abnormal growth responses.

143

Specific groups of damage symptoms have been characterized at specific locations in Germany (FBW, 1986). These include the yellowing of needles of spruce trees growing both in the low mountain ranges and at high elevations of the limestone Alps; distinct dieback and thinning out of tree tops at medium and higher elevations, as in the Harz Mountains and in forests near the coast; and reddening of needles of older tree stands growing in southern Germany. These may overlap so that geographical zones cannot be clearly delimited. An excellent detailed description is provided by Professor Robert Guderian and his colleagues (Guderian *et al.*, 1985).

In Europe the new decline symptoms first became apparent on silver fir (*Abies alba* Mill.) in the early 1970s, but fir comprised only about 2–4% of the forest in the Federal Republic of Germany. The Norway spruce (*Picea abies* (L.) Karst), which comprised over 40% of the forest and is the most commercially important timber species, remained healthy. Then, in 1976, symptoms appeared on spruce in Norway and Sweden as well as in Germany. By 1979 damage reached serious proportions in central Europe from France to Czechoslovakia and Poland, where it became most severe. Since 1980 symptoms have appeared on Scotch pine (*Pinus sylvestris* L.), beech (*Fagus sylvatica* L.) and oak (*Quercus robur* and *Q. petraea*). Did this decline have the same origins?

Symptoms appeared rather randomly, first and most commonly on trees over 60 years of age at elevations above 600 m (1969 feet) near both sides of the Czechoslovakian border and in the Black Forest (Prinz *et al.*, 1982). By 1984 symptoms were reported to have appeared not only in 50% of the forests throughout Germany, but over much of Europe as well (Federal Ministry of Food, Agriculture and Forestry, 1984; Nilsson & Duinker, 1987). All types of damage were recorded in the 1984 survey, including more familiar fungus and insect-related diseases as well as symptoms of the recent decline, often in its mildest expression. (Prinz, 1985; Schopfer & Hradetzky, 1986). Affected species now included beech, birch, maple, ash, alder and oak, as well as the conifers. The situation had not improved in 1986, although the rate of the decline slowed markedly. Still, more than 14.2 million hectares (35 million acres) of forests in Europe were damaged to some degree by the recent forest decline (Deumling, 1987) according to the new classifications.

Descriptions of decline often neglect the most critical aspect, which is the decreased radial growth of affected trees. Before the appearance of visible foliar symptoms, trees respond first to stress by a suppression in growth. Whatever the cause of decline, radial growth reductions are an early indicator of stress. These reductions are reflected in the narrower tree rings that are detectable by examining either tree cross-sections or cores taken from the trunk. Dendrochronological techniques reveal not only the amount of annual growth but the years in which suppression occurred. But dendrochronology does not directly reveal the cause of the suppression. For this, related techniques must come into play and much of this relies on other observations and sound deductive reasoning.

GROWTH RESPONSES

The first step in evaluating alternative hypotheses explaining the cause of decline is to provide a detailed characterization of the timing and nature of alterations in the historical growth associated with decline. When growth ring analysis reveals when decline first appeared, we can relate this to any stress events that occurred at that time.

Dendrochronological studies in the eastern USA are fairly consistent as to the timing of periods of decline, although conclusions as to the causes vary. One thorough study in the Adirondack Mountains of New York reflects the general growth trends in that area (LeBlanc *et al.*, 1987). Decline in red spruce growth was characterized by a loss of growth potential beginning from 1964 to 1966. A second period of decline occurred between 1978 and 1983. The mid-1960s was also the time of the most severe regional drought in 240 years, while the later period coincided with periods of abnormal winter temperatures. Exceptionally warm winters occurred in 1980–81 and 1982–83. Consecutive unusually cold winters appeared from 1976 to 1979. Injury to terminal buds was postulated that was thought to affect overall growth. Another hypothesis held that excess atmospheric nitrogen inputs stimulated apical growth to continue late in the season, predisposing trees to winter injury. Excess nitrogen indirectly also may decrease root growth, making trees more susceptible to drought. However, the coincidental appearance of non-climatic factors must not be overlooked.

Even where ring width changes are not pronounced, wood density may be. Density is a function of cell size and extent of secondary cell wall thickening. Density normally correlates well with climate, especially spring temperatures. For red spruce in Maine, where ring widths showed no notable differences, the correlation of wood density with climate failed to hold after the mid-1960s (Conkey, 1987). No clear reason appeared for this unprecedented change in relationship between density and climate, but the increasing intensity of air pollution may have played a significant role.

Growth reductions also have been demonstrated in Europe, although the timing is not always the same. Dendrochronological studies have been most frequently applied in Switzerland where, in Bremergarten, Aargau, growth reductions can be dated as beginning in the 1940s or 1950s (Schweingruber, 1986). Some growth reductions of fir and spruce seem to have been initiated in very dry years, while others began in very wet years. Exceptions occur near industrial sites, where growth reductions are more closely related to pollution.

Other studies in Switzerland suggest an interaction of pollutant emissions with climate. Growth of silver fir, Norway spruce and Scotch pine during the past century has been poor in dry years (Kontic *et al.*, 1987). But there are exceptions. There was severe suppression of silver fir growth in 1973, during a period of favorable climatic conditions. Spruce growth reduction generally was closely correlated with dry years, but there was little suppression before 1950, and the proportion of damaged trees increased after the early 1960s. The authors believed that the only plausible explanations for damage was pollution stress. Drought may have triggered the suppression, but in the area studied the correlation was strongest with industrial development, including the local aluminum industry.

Growth increment reduction in Germany also began before visible decline symptoms became evident. Raw ring widths in one study were markedly reduced in the drought year of 1976, but decreases also were noted in the 1950s (Bauch, 1983) and were not always associated with dry years. Other research indicates that reduced growth of Norway spruce and silver fir in some cases began in the 1960s (Von Eckstein *et al.*, 1983).

Tree ring measurements of *Abies cephalonica* near Athens, Greece, showed a decline in growth beginning in the 1960s. This correlated with the rapid urbanization of Athens, together with an annual increase of 15% in the numbers of automobiles. Air pollution was suspected but no monitoring data were available (Heliotis *et al.*, 1988).

The underlying mechanisms explaining growth anomalies may lie in carbon partitioning. Research in the Black Forest has shown, for instance, that the ATP/ADP ratio is altered, with a higher ratio appearing in trees showing decline symptoms. Fructose 2,6-bisphosphate is also lower in symptomatic trees. There is also a pronounced reduction in NADP in the injured trees and a higher NADPH/NADP ratio, showing changes in energy relations and an increased rate of senescence (Magel *et al.*, 1988). Ozone concentrations in the study area averaged an elevated 170 $\mu g/m^3$ (87 ppb) during the summer, double the generally accepted background.

WHAT CAUSES DECLINE?

A wealth of clues is revealed in the symptoms and overall syndrome of decline that should have led to an obvious diagnosis and solution to the phenomenon. But many are contradictory, and progress has been impeded by numerous obstacles. An objective look at decline provides a striking example of how the communications media and politics have influenced the thinking and study of a major disease complex.

Looking first at the process of diagnosis, one seeks a common thread which binds all the proposed causes of a disease—in this case decline. Did this decline involve a common species, a common soil type, similar atmospheric or climatic conditions, consistent presence of some pathogenic organism, or any other stress? With the recent decline no obvious commonality is apparent. The crucial question still looms: Why did decline develop relatively rapidly and over such a broad geographical area so diverse in elevation, environment and numbers of species affected?

It is unlikely that any one of the hypotheses discussed provides a single answer to decline. It is even impossible to rank the hypotheses in order of probability. Certain factors, such as climate, perhaps in combination with ozone, may have weakened the trees and predisposed them to other stresses. Once a tree has begun to decline, contributing factors, often biotic stresses, may ultimately accelerate the decline (Manion, 1981). In reviewing the hypotheses it is important to keep in mind that two or more pathogens (i.e. stress factors) may be acting in concert to cause decline, and that these may not always be the same in different geographical regions (Prinz, 1987).

Climatic factors

From a logical perspective, once the obvious known pathogens have been eliminated as a cause, one might reasonably next look to unusual weather conditions. Sudden drops in temperature, warm periods in winter, drought, ice storms or hurricanes bringing in salt from the sea, all can impair plant health (Treshow, 1970). The widespread occurrence of decline could implicate some global or regional weather phenomenon such as drought, which might at least serve as a 'triggering' mechanism that predisposed trees to secondary stresses. Plant injury in general is particularly intense following dry years, winters with low temperatures, or late frosts (Wagner, 1981). Did such events occur synchronously with the appearance of decline?

An examination of weather conditions revealed that drought conditions were pervasive over much of central Europe in the 1970s and early 1980s (US Department of Commerce, 1986).

In the city of Kassel, near the Harz Mountains of Germany, where decline has been especially severe, every month of 1976 except January recorded well below normal amounts of precipitation. The summer months from May to September had only 57% of normal amounts. Drought was still more critical in 1982. During the 5-month summer period Kassel normally could expect to receive 321 mm (12.6 inches) of rain. In 1982 it received only 182 mm. The critical months of July, August and September received only 65 mm: 129 mm less than normal. Except for 1977, the intervening years all registered precipitation amounts well below normal.

Kassel was not unique. Similar conditions prevailed over much of Germany and Europe in general. In Stuttgart, near the Black Forest, which is another area of serious forest decline, 1976 precipitation was only 76% of normal. It had also been dry in 1975 and during every subsequent summer until 1983, except for 1977. July to September, 1979, was the most critical period, having received only 129 mm of the expected 217 mm of precipitation. The consequences of low summer rainfall over a period of years, in forests accustomed to much more, might well have been expected. The effects are reflected clearly in the observed reduced increment growth (Schopfer & Hradetzky, 1986).

The same conclusion was reached in Sweden and Norway, where reductions in vitality and crown density of Norway spruce and Scotch pine were increasingly severe as one travelled north. Climatic stress and aging effects overshadowed possible effects of pollutants (Tveite, 1984). The shoot dieback that occurred on pine and spruce trees in Norway in 1974 had been attributed largely to water stress in the affected areas (Braekke, 1980).

The appearance of recent decline over a broad area and subsequently on many different species strongly suggests that a non-specific causal agent such as drought could be involved. But some believe drought was not a factor, or did not act alone, and the issue remains controversial (Eckstein & Bauch, 1983).

Frost shocks in late winter or early spring in Germany after long warm periods as occurred in 1981, 1982 and 1983, could have provided stress, contributing further to the decline (Rehfuess, 1985). Winter damage to foliage, possibly predisposed by other stresses such as gaseous pollutants, may have contributed still further (Friedland *et al.*, 1984).

The most insightful reviews of the possible role of climate in causing forest decline were presented in a conference held in Munich in 1986 (Kirschner, 1987). While the cause was not established, frost shocks were strongly implicated. Specifically, Kirschner noted that 'growth reductions seem to have been initiated in very dry years'. These were notably 1921, 1944 and 1976. The May to August periods were most critical. Very cold, wet years, most significantly 1948, also were related to growth reductions. This and any similar frost shocks would, at least, have an underlying influence on the severity of decline the following summer.

Ten years of abnormally dry summers were also thought to have been a major contributing factor to such secondary stress factors as bark beetle infestation and pathogenic root fungi, as well as to decline on the whole. The strongest climate–growth decline relationship occurs when hot summers are followed by low December temperatures (Prinz, 1987; Johnson *et al.*, 1986).

Fritz Schweingruber at the Swiss Federal Institute of Forestry Research (1986) states that 'the frequency of growth reductions is governed by climatic conditions and other, unidentifiable factors. Only in a few areas is pollution the probable cause'.

Although drought, or frost shocks, alone could have been critical in some instances, the consequences of one or both acting in concert with other pathogens also would weaken trees further, allowing fungi, such as those causing needle casts, to become especially active in causing premature shedding and thinning of foliage. But none of these stresses alone account for the totality and diversity of symptoms observed. For this, secondary factors or interacting pathogens, biotic or abiotic, must be considered (Halbwachs, 1988).

'Acid rain' and fog

Forest effects

Many far-ranging hypotheses have been proposed to explain both the new forest decline phenomenon and the possible component of atmospheric pollution that might be a causative factor (McLaughlin, 1985; Schutt & Cowling, 1985). It was 'acid rain' that caught the media attention and the public fancy. The term 'acid rain' actually was first used by Robert Angus Smith in 1872 (Wellburn, 1988) to describe the rain falling around Manchester, England. The concept initially proposed that acids, such as sulfuric and nitric acids, formed from pollutants in the atmosphere and made rains and snows abnormally acidic. But the term has come to infer all types of long-range transported pollutants that might occur in forest ecosystems.

The changing acidity of air and precipitation over North America and much of Europe was noted some years ago (Oden & Ahl, 1970), at which time the concern centered on the threat to aquatic ecosystems. Similar risks were noted in the USA (Likens & Borman, 1974), and potential effects of acid precipitation on vegetation were pointed out (Oden, 1976; Tamm & Cowling, 1976). Large-scale acidification of forest soils in Germany were recorded between the 1920s and 1980s (FBW, 1986; Tamm & Hallbacken, 1988). However, rain in Germany had already reached a point of 'acid saturation' by 1940 when the pH value reached 4.2, making further acidification and adverse responses unlikely (Winkler, 1982). It should be noted that rain water is not neutral (pH 7) but normally has a pH of about 5.6.

When forest decline appeared in Germany, Bernard Ulrich and his colleagues (1979), who were already studying aluminum toxicity and nutrient cycling, proposed that the acidification of soils was accelerated as a result of deposition of acidifying substances from the atmosphere. Increased acidity, it was proposed, made aluminum and many other ions in the soil more soluble and, hence, more mobile. The greater concentrations of aluminum in the soil were considered to be toxic and damaging to root systems. They hypothesized that direct aluminum toxicity led to a reduction in the fine roots of the trees, thus ultimately weakening and killing the tree (Ulrich *et al.*, 1980).

Although it was the most publicized theory, and stimulated tremendous research interest, there were some problems. Symptoms of aluminum toxicity did not closely resemble those of forest decline; decline was severe in calcareous and alkaline soils that were very low in aluminum, and, most disconcerting to the hypothesis, forests suffering the most 'severe decline were not always recipients of the greatest acidity. There was no correlation between aluminum content in soils and the intensity of needle chlorosis. Eventually researchers came to believe that if aluminum were involved at all, it was most likely in its ratio to calcium or magnesium.

In addition to aluminum, acidity could mobilize, i.e. increase the solubility, of other elements, including manganese and other heavy metals. Cadmium, lead, copper, manganese, arsenic and zinc deposition are all far higher in central European sites than in rural US locations (Lindberg *et al.*, 1985; Brechtel *et al.*, 1986). Garbage burning, coal burning, iron and non-ferrous metal production and the glass industry all contribute to atmospheric deposition of cadmium. Mobilization caused by soil acidification adds to the availability of the chemicals already present. The concentrations of zinc in the soil solution of the surface humus are as high as those that stunted root growth in nutrient solutions. Generally, though, foliar concentrations of these elements are low enough that this possibility tends to be discounted (Prinz, 1985). While these might be toxic at the higher concentrations, once soluble they are more subject to leaching. Many plants are adapted to high concentrations of such elements (Rehfuess, 1981), and plants may show decline symptoms whether or not these elements occur in excess.

Acidification might affect soils in other ways. But soils are extremely well-buffered systems, and an increased input of hydrogen ions or acidifying substances does not translate directly to a change in soil pH or acidity. Nevertheless, evidence for enhanced soil acidification comes from measurements of old and new soil samples, geochemical studies on soil profiles, and input–output budgets of cations and anions in soils and watersheds (van Breeman & Mulder, 1986). As a result, mineral acidity has largely replaced organic acidity as the dominant acid source in forest soils. Should the pH of throughfall water fall below 3.6–3.8, mycorrhizal development may be impaired (Dighton & Skeffington, 1987). Rainfall pHs in this range are characteristic in many forests.

Other potential effects that might occur have been postulated in relatively unbuffered, inorganic soils. These have involved largely an altered availability of nutrients (Abrahamsen, 1980, 1984). Saturation of ecosystems with atmospheric nitrogen from oxides of nitrogen and ammonia increases the probability of further soil acidification. Other effects of acidity involve the increased leaching of calcium, magnesium and other cations from the upper soil horizon, or reduced microbial decomposition and release of nutrients from the organic matter.

A major Norwegian project conducted between 1972 and 1980 intensively investigated the effects of acid precipitation. The study concluded that 'forest growth effects due to acid deposits have not been demonstrated' (Overrein *et al.*, 1981). Recent studies in Canada also disclosed no relation between acid deposition and growth, specifically of Norway spruce, red pine and Scotch pine (LeBlanc *et al.*, 1987).

A recent annual report of the National Acid Precipitation Assessment Program (1986) draws some rather startling conclusions. Namely, 'forests are probably relatively unaffected by ambient acidity in rain on their foliage at low elevations or even in the mist of above-cloud-base forests'. Simulated rain at average pH of 4.2 and average acidity of the clouds at mostly pH 3.6 showed no negative effects on seedling germination or growth. On the other hand, the report noted that 'forests in the U.S. are probably stressed to some extent by ambient ozone levels.'

Crop effects

If acid rain, or deposition, has a potential to injure forests, crops also are likely to be affected. This prospect has been intensively studied in Europe and North America since the 1970s (Ferenbaugh, 1976; Shriner *et al.*, 1977; Irving, 1987; Pell *et al.*, 1987), and less intensively much earlier. Annual crops could provide almost immediate answers, even though any negative findings do not preclude the possibility of long-term effects on perennial crops or trees.

A nationwide program to monitor acid deposition began in 1978 with 14 National Atmospheric Deposition Program (NADP) sites. It has been expanded to include all the states, often with many sites per state. Similar programs function throughout Europe. Data from these networks now make it possible to answer the basic question, 'What is the acidity of the deposition?'

Much of the research effort has addressed the possible influence of simulated acid rain applied at realistic concentrations throughout the growing season. The principal question asked was, 'Is yield affected by acidity?'

Numerous studies address this question, which is succinctly answered in the NAPAP Annual Report to the President and Congress (1986). Reliable studies had been completed for eight important crop species exposed to simulated acidic rain. These indicated 'no consistent effects from ambient levels of acidic rain' (pH 3.8–5.1).

A review of 90 studies involving over 30 plant species or cultivars disclosed that growth could be stimulated or inhibited depending on the pH. Above pH 3.5 most of the responses were either positive or absent. Below this, responses were mostly negative (Table 12.1) (Treshow, 1983).

Collectively, these and more recent studies suggest that acid deposition has little or no direct role in affecting crop production (Irving, 1987; Elliott *et al.*, 1987; Pell *et al.*, 1987). The greatest direct risk would be in the necrosis that can occur when rain pH drops below about 3.0. However, the NADP data indicate that this is extremely infrequent.

Table 12.1. Plant growth and production responses to simulated acid deposition[1]

pH value	Response[2]		
	Positive	Zero	Negative
3.5 and over	14	71	7
3.0–3.4	9	39	11
2.5–2.9	4	2	6
2.0–2.4	0	2	0

[1] Numbers refer to the frequency of individual studies.
[2] Response: 'Positive' refers to production greater than the controls; 'negative', less.

Acidic fog

The acidity and composition of fog also should be considered. Compared to rainwater in general, trace element concentrations in fog water show increases by factors of 10–15 (Schmitt, 1986). Concentrations of ammonium ion were enhanced by a factor of 14. The pH of fog tends to be slightly lower than the pH of rainwater. One study noted values as low as 2.5, but the mean value was 4.2 (Schmitt, 1986).

The fundamental mechanisms of acidic fog presumably are much the same as for acid rain. In either case we must bear in mind that the plant cell is a remarkably well-buffered system. The high carbonate and phosphate content especially would infer that direct chemical changes from acidic input would be unlikely except under extreme conditions. Any acidification effects are more likely to be indirect. However, exposure studies have shown that leaf injury can be produced when fog pH is below 3.0, although most plants studied were injured only below pH 2.0 (Musselman & Sterrett, 1988).

Nutrient relations

Excess nutrients

The addition of nitrate (NO_3^-) and sulphate (SO_4^{2-}) to soils, increasing their fertility, will affect plant health (Ellenberg, 1984). Nitrogen enrichment by oxides of nitrogen and ammonia through foliage is also a possibility. Ammonia supplements have been postulated to provide an additional explanation of forest decline in Europe (Nihlgard, 1985). Trees can absorb and utilize many essential elements through the foliage as well as the roots, and excess ammonia has been found to cause defoliation of spruce after 3–4 years (G.W. Heil, 1988, personal communication; Van der Ferden, 1982).

Nitrogen is the element most often limiting productivity, so this supplement theoretically might be beneficial. However, in excess, nitrogen has the adverse effect of adding to the demand for other elements that might become in short supply relative to the nitrogen. Excess nitrogen especially may cause deficiencies of magnesium, potassium, phosphate, molybdenum and boron. These deficiencies may impair the assimilation of nitrogen and the synthesis of proteins. Excesses also could lead to the formation of larger, thin-walled cells—a 'succulent' type of growth less tolerant to stress and predisposing the trees to frost or fungus pathogens. The oversupply of nitrate and undersupply of carbohydrate may also adversely affect symbiotic mycorrhizal fungi (Bjorkman, 1949). Given a high nitrogen supply, root tips grow out of the mycorrhizae. They become exposed and grow freely and may be damaged if the soil solution contains metal ions or heavy metals.

Nutrient imbalances do not provide a well-substantiated explanation for decline. The nitrogen excesses in the USA are at least as severe as those reported in Europe. Yet, in the USA, the decline syndrome is different. Conifer needle yellowing and twig dieback are not the same, and few species are affected. Decline in the USA appears to be a different disease, and nitrogen nutrition does not appear to be involved.

Nutrient deficiencies

A deficiency of plant nutrients has been hypothesized as a possible cause of forest decline, if not alone then in association with other stresses. The most logical of associated interactions lies in the occurrence of prolonged dry periods early in the growing season when demands for critical nutrients are greatest. The limitation of available water in which the nutrients could be transported into the tree could seriously limit uptake and jeopardize the trees' survival.

Nutrient deficiencies, most notably of magnesium and calcium, are of greatest concern. Leaching from foliage, aggravated by acidified rain disrupting the cuticle and cell membranes, is further intensified by gaseous pollutants (Rehfuess, 1983; Zoettl & Huettl, 1986). Magnesium was considered to be especially sensitive to leaching. In the Bavarian Forest magnesium concentrations in needles of declining spruce trees growing at altitudes above 1000 m have been found to be extremely low (Bosch *et al.*, 1983). The trees exhibited the characteristic deficiency symptoms of yellowing and premature shedding of older needles and of growth depression. The deficiency would lead to decreased frost hardiness of the needles, thereby enhancing decline. Magnesium deficiency symptoms are not prominent in all decline areas, however, so other causes of decline must be sought there.

Declining trees also can be characterized by symptoms of boron deficiency. In general, boron deficiency causes yellowing of needles, premature shedding and general decline. More specific symptoms, including the gumming and cracking of stems, the presence of weakened, drooping branches and the proliferation of adventitious shoots, as is found in some cases of decline, are almost unique to boron deficiency (Treshow & Deumling, 1985). The involvement of boron with growth regulators also provides a strong clue that it might be associated with decline. For example, trees showing decline symptoms have increased concentrations of abscissic acid, indicating a perturbation in hormone relations, which is another response to boron deficiency or imbalance (Schutt, 1983). Increased sensitivity to frost also characterizes boron deficiency.

Light-textured soils that are low in organic matter, as occur in the high-elevation, granitic soils of the Bavarian Alps, are most sensitive to boron deficiency problems (Russell, 1973). This further supports the hypothesis that boron deficiency, triggered by drought and possibly exacerbated by elevated nitrogen, may be responsible for some of the decline observed. If boron were involved in decline, possibly through boron/nitrogen imbalance, it would help explain how so many different species could be affected. Also, older trees, which are most commonly observed to show decline, are also most sensitive to boron deficiency. Another argument is that boron is primarily involved in meristem areas and interacts with hormonal mechanisms. Disruption of these processes have been the most difficult to explain with other hypotheses. It might also be noted that decline has never been observed on agricultural crops. Since these are generally fertilized, and boron is thus provided at least as a trace element, the boron deficiency hypothesis would explain these phenomena (Treshow & Deumling, 1985).

Boron deficiency, with severe shoot dieback, was especially striking in Norway in the low rainfall years of 1974–1976 (Braekke, 1979). Research in the Scandinavian countries (Kolari, 1983) further supports the importance of boron to decline. Research in the German Democratic Republic (Fiedler & Hohne, 1984) reported the occurrence of normal baseline boron concentrations and ratios with other nutrients but did not address the decline problem. Spruce and pine required 20–45 ppm boron; and concentrations in organic soils below 16 ppm were considered deficient. However, only a single study of needle tissue analysis is available (Rehfuess, 1985). This showed concentrations of 6–18 ppm boron in healthy as well as declining spruce trees in southern Bavaria. Rather than the boron concentrations alone being critical, it is more likely that the boron to nitrogen ratio as well as the relative amounts of calcium are most significant.

Biotic pathogens

Fungus diseases, including those caused by root- or wood-rotting fungi, generally do not appear to cause the new decline, other than by attacking predisposed trees. One group of biotic pathogens has been implicated. These are the needle cast fungi (Rehfuess, 1985). In the late autumn of 1982 and

1983 older Norway spruce stands over southern Germany were heavily attacked by *Lophodermium piceae*, *L. macrosporum*, *L. filiforme* and *Rhizosphaera kalkoffii*. These fungi, while often considered to be secondary pathogens (Butin & Wagner, 1985), can cause premature shedding of older needles, rendering the crowns more open and transparent (Rehfuess & Rodenkirken, 1984). Such trees may be predisposed to infection following frost shocks or other stresses.

Gaseous pollutants

Sulfur dioxide

No one questions that sulfur dioxide can cause classic forest decline. This was established over a century ago (Sorauer, 1886; Hasselhoff & Lindau, 1903). The question is, does it cause, or is it implicated in the recent decline? When this phenomenon first appeared in Bavaria near the Czechoslovakian border, sulfur dioxide was considered to be responsible as it had been in the CSSR for several years. Jan Materna (1984) at the Institute of Forestry reviewed studies showing that dieback, decline and predisposition to frost of silver fir occurred when atmospheric sulfur dioxide concentrations averaged about 20 $\mu g/m^3$ (8 ppb) over an 8-year period; Norway spruce was damaged by 50–70 $\mu g/m^3$ (19–27 ppb). Average annual concentrations of sulfur dioxide in northeastern Bavaria are in this range, and total sulfur in the needles approach the elevated 2% (Rehfuess, 1985). Damage was most severe at higher elevations near the GDR–CSSR border, as well as in Poland, and it is probable that sulfur dioxide is involved in decline in those areas. In areas including the Bavarian Forest and Black Forest, direct effects of sulfur dioxide are less likely since average annual ambient concentrations are only in the range of 10–36 $\mu g/m^3$ (4–14 ppb) (Prinz et al., 1982). However, observed peaks up to 300 $\mu g/m^3$ (114 ppb) raise some questions (Baumbach, 1986). Since 1970, however, sulfur dioxide concentrations have decreased while the severity of decline has increased.

Where classic decline symptoms are most prevalent in Poland, Czechoslovakia, and the GDR, notably the Erzgebirge, sulfur dioxide may be solely responsible. But in other instances, the effects of sulfur dioxide may be more insidious than direct. In addition to the accumulation of sulfur in the foliage, there is a considerable build-up in the soil and groundwater. At least as seriously, sulfur dioxide disrupts permeability of the cells, allowing other elements, most notably magnesium and calcium, to wash out of the cells and to leach readily through the canopy (Seufert & Arndt, 1986). When ozone is also present the impact on nutrient loss is greatly enhanced. The leaching of manganese can be doubled.

It is important to note that more than the sulfur dioxide concentrations are involved. The annual deposition can be important, not only of sulfur dioxide but also other chemicals (Deumling, 1986).

Ozone

It was 1982 before ozone was suggested seriously as a cause of forest decline in Europe (Arndt *et al.*, 1982; Prinz *et al.*, 1982). Despite the well-known history of ozone effects on vegetation in the USA, ozone did not have the immediate media appeal of acid rain. Yet elevated concentrations of ozone have been known to occur over much of Europe since the 1960s (Darley, 1963/64; Knabe *et al.*, 1973). In the 1980s, mean summer ozone concentrations in many areas were about 50 ppb (98 $\mu g/m^3$), and concentrations above 100 ppb occur in 3–5% of the summer hours (Ashmore *et al.*, 1985). These concentrations are roughly above the damage threshold for sensitive annual crops, but little is yet known regarding the consequences to perennial plants when such exposures are repeated year in and year out.

At higher altitudes, above 600–800 m, decline seems to be greatest and so are ozone concentrations. In one German study in 1976 (Arndt *et al.*, 1982), concentrations near Bonn were compared with concentrations at rural mountain sites south of Bonn. The highest urban concentrations were 60–100 ppb, whereas the mountain site concentrations reached 120 ppb. But, most critically, concentrations never fell below 50 ppb in the mountains. This and other studies have shown the monthly and annual mean concentrations consistently to be higher in the mountains (Prinz *et al.*, 1982). Regional maps show that ozone concentrations exceed 75 ppb 100–200 times each year in the southern half of Germany as well as in parts of France and Austria. This is also the area of the most serious decline (Hauhs & Wright, 1986). Ozone concentrations here are similar to those reported in the eastern USA (Stasiuk & Goffrey, 1974; Skelly *et al.*, 1983) and recent observations in the Rocky Mountains. Ozone concentrations near Athens, Greece, exceed 120 ppb during pollution episodes and approach 200 ppb, but forest decline does not yet appear to be a problem (Gusten *et al.*, 1988).

Symptoms caused by ozone can be similar to those characterizing the new decline. The basic subliminal or 'hidden' symptom of photosynthesis suppression is common to both. Also, ozone accelerates senescence of the older needles, causing premature needle cast and canopy thinning as well as causing visible symptoms on sensitive individual trees. One problem in considering ozone, however, and in conducting ozone fumigation studies, is that there is considerable variation in sensitivity among individuals. Hence, a group of trees selected to be fumigated may be relatively tolerant to ozone. The trees may not necessarily be representative of the field population. Often no visible symptoms are produced in controlled ozone exposures even at high concentrations (Prinz, 1983). The classic chlorotic mottle has been produced on Scotch pine, but only at prolonged mean concentrations of 100 ppb for 56 days (Skeffington & Roberts, 1985). Fine root biomass was reduced, and older needles became senescent at 150 ppb. Spruce, fir and beech proved to be less sensitive.

Another problem in attributing decline to ozone is that not all of the symptoms of decline are accounted for by ozone. The needle mottle and tipburn common in the USA are usually absent in Europe. Also, the gumming and cracking of bark, adventitious shoot growth, and weak, drooping branches have never been associated with ozone. Furthermore, the nutrient deficiencies often associated with decline in Europe do not occur in the USA, where ozone is as major factor but where nutrient deficiencies have never been associated with ozone.

Such inconsistencies have led some researchers to consider the combined effects of ozone with sulfur dioxide, acid deposition or other stresses. Numerous studies on a number of forest and agricultural species have addressed this possibility. Although most studies in the USA have yielded negative results, some studies in Europe have shown that interaction can occur (Seufort & Arndt, 1986; Guderian *et al.*, 1988). Perhaps most significantly, the combined effects of ozone and acid precipitation negatively influence the nutrient budget of plants (Prinz *et al.*, 1984).

The broad range of symptoms and diversity of species showing decline in Europe are not reflected in US forests. Critical examination of the syndrome in the two areas indicates that they are two distinct phenomena. Similarities of some of the symptoms may be shared, but the syndrome, i.e. the composite of all symptoms and interactions, is different.

The best-documented decline episodes in the USA involve the appearance of visible symptoms. In the west, ponderosa pine (*Pinus ponderosa*) trees show characteristic chlorotic mottle; in the east, white pine (*P. strobus*) shows either the mottle or tip necrosis characteristic of ozone toxicity. Symptoms on spruce (*Picea*) are more insidious. There is a gradual weakening of the trees, followed by dieback and ultimately death. The mottle symptoms described on ozone-fumigated spruce (Treshow, 1970; Krause, 1986) have not been reported to occur in US spruce or fir forests. Attention in the 1980s has focused on possible growth reductions of several species, most notably pine and spruce.

In the eastern USA, the first abnormal decline in trunk radial increment width was reported on pitch pine (*Pinus rigida*) and shortleaf pine (*Pinus echinata*) to have begun in 1957 (Johnson *et al.*, 1981, 1983). Perhaps coincidentally, this is the same year in which a number of diseases were found to have been caused by ozone in the same area (Hill *et al.*, 1961). Decreases in increment growth of red spruce (*Picea rubra*) began in the mid-1960s in New York and New England (LeBlanc *et al.*, 1987). Since this period coincided with severe drought in the same areas, was ozone or drought the most critical? Studies conducted in the Appalachian Mountains from Virginia to Tennessee in the 1970s indicated that reduced increment growth in that area also began in the 1960s (Skelly *et al.*, 1983). In this instance, however, the study species was eastern white pine and it exhibited symptoms typical of ozone toxicity; the increment growth decreases were attributed to this pollutant. Subsequent studies further suggested ozone to have caused the reduction (Benoit *et al.*, 1985). Mean monthly ozone concentrations in the study area were roughly 20–50 ppb, with peaks often exceeding 100 ppb.

In a dendrochronological study in Maine, tree-ring indexes were calculated from increment measurements on white pine trees showing injury symptoms attributed to ozone and on similar, asymptomatic trees in the same stand (Treshow *et al.*, 1986). Trees showed characteristic ozone symptoms in 1984 but apparently none before that year. Tree-ring indexes of symptomatic trees were significantly less than in the asymptomatic trees after 1978. The study suggested that subliminal effects of ozone-induced growth suppression began in southeast Maine by 1979. Exceedences of the Maine State Air Quality standard for ozone of 80 ppb were considered to be the most accurate measure of potential plant response. There were 101 and 88 such exceedences in 1983 and 1984, respectively.

Growth reductions have been documented in the Sierra Nevada Mountains of California (Peterson *et al.*, 1987). A mean reduction of 11% in growth indexes since 1965 was reported on Jeffrey pine (*Pinus jeffreyi*) trees that showed symptoms of ozone injury.

The National Acid Precipitation Assessment Program (1986) reports that coniferous seedlings exposed to ozone in the ambient concentration ranges found in forested areas of the USA showed a decrease in growth with increasing ozone concentrations over the range of 20–100 ppb for all species tested. When the growing season average exceeded 50 ppb (based on daily 7-hour concentrations), significant growth reductions of a few per cent per year were noted. Visible symptoms occurred following short-term peaks in the 100–200 ppb range. Direct mortality was found when 200 ppb was exceeded 10–20 days during the growing season, as in southern California.

It must be emphasized that growth reductions in part of a population need not constitute a decline of the forest, certainly not the type of decline that is so serious in Europe. While ozone can cause growth reductions, damage, and even visible symptoms on many sensitive individuals, it is too early to establish ozone as the cause of forest decline, with the possible exception of decline in California.

There is evidence suggesting that ozone may delay the normal hardening processes that prepares spruce and other plants for winter. Such a failure of the leaf cells to develop protection against the cold would be especially serious in years of sudden temperature drops. The subliminal adverse histological effects on cell ultrastructure might make the cells still more sensitive to low temperatures, as is the case with sulfur dioxide (Huttunen, 1984).

The evidence that ozone is responsible for the new forest decline is plausible but circumstantial. Ozone might largely explain forest decline in some areas, but not all. Where ozone is not a primary cause of decline, it still might act to predispose trees to other stresses, such as frost, or it is possible that ozone might interact with other stresses. Ozone is especially important in damaging the leaf cuticle, thereby enhancing the washing out of cell cations such as magnesium (Karhu & Huttenen, 1986). Directly or indirectly, ozone appears intimately associated with forest decline in at least some areas (Woodman & Cowling, 1987).

Hydrogen peroxide

Forest decline in central Europe seems to be concentrated in areas most frequently exposed to fog or intercepted rainwater (Prinz *et al.*, 1982). It has been proposed that the hydrogen peroxide in cloud and rainwater, whose concentrations range from 0.1 to 5.0 ppm (Kok, 1980; Masuch *et al.*, 1985), might be important in causing decline. Ozone is a highly reactive molecule and in solution may react with unsaturated fatty acids, sulfhydryl and ring-containing compounds. Hydrogen peroxide may be formed by decomposition reaction of ozone with olefinic double bonds in solution, which produces an aldehyde or ketone and a Criegee zwitterion (Tingey & Taylor, 1982). In water, ozone decomposition products also include hydroxyl, hydroperoxyl, superoxide anion and other free radicals that may be more reactive than the ozone *per se* (1976).

Hydrogen peroxide thus provides one more hypothesis; one that, to date, has been the least examined. Each hypothesis, in turn, appeared promising and each had its advocates. Hydrogen peroxide has been studied too little even to assess its potential involvement in decline. The extent to which this avenue is pursued will depend to a large extent on the continued interest of researchers, public concern and political attitudes, and, of course, on whether the decline itself persists.

CRITICAL APPRAISAL

The recent or 'novel' forest decline is a critical problem threatening forests in the industrial nations. Many of the declines can be explained to result from well-known pathogens, but many more cannot. There is still controversy as to which are well-established 'natural' phenomena and which fall into the category of 'Neuartige Waldschaden'. There still exists a question as to whether the recent decline in different geographical regions is caused by a single stress, or by a combination of factors. Also, are the symptoms and the causal pathogens the same in different regions? A general consensus is beginning to emerge.

First of all, there are two or more distinct types of decline, each having a separate etiology. The major cause is complex, with many of the stresses acting sequentially or collectively to cause decline. But it is not yet known which stresses are the most critical.

Different symptoms appear in different geographical regions. For instance, the situation in North America and some of the European forests is clearly distinct. Also, as reviewed by Guderian *et al.* (1985), symptoms in different parts of Germany are different. For instance, in northern Germany, injury usually occurs in the crowns of dominant trees that rise above the general canopy level. Needle casting and yellowing typically begin with the 3–4-year-old and younger needles towards the upper part of the crown (sub-top-dying). This is most characteristic of air pollution. In southern Germany thinning of the crown proceeds from the bottom to the top and interior to exterior, and branches show more drooping. Gumming and adventitious growth also are more in evidence in these forests. Widespread yellowing of all needle ages is conspicuous, suggesting the involvement of adverse nutrient relations. Other symptom patterns also are evident, as where forests are affected by sulfur dioxide.

The 'Komplexkrankeit' hypothesis, formulated by Schutt *et al.* (1983), postulates that the total impact of air pollutants in the past decades, and their combination, leads to a successive reduction in carbohydrate production of forest plants. This results in severe loss of vitality and an increase in predisposition to plant pathogens, including abiotic stresses such as frost and drought. It does not account for the growth abnormalities or symptoms on different species.

The available data best support the thesis that the new type of forest decline is directly or indirectly associated with ozone in some areas and possibly sulfur dioxide in others. Ozone in combination with sulfur dioxide and nitrogen dioxide appears to be especially harmful (Guderian *et al.*, 1988). Disturbances in nutrient relations, brought on by nitrogen excesses or imbalances, and perhaps a succession of dry summers creating chronic moisture stress, are also critical. This may lead to magnesium and possibly boron deficiencies, especially when coupled with excess leaching aggravated by loss of integrity and erosion of the leaf cutin and waxes caused by the presence of ozone or acid deposition. Soil characteristics also have a major role in metabolic changes within the plant and the plant response to pollutants and other stresses. As such, the soil is important to manifestation of damage symptoms.

SUGGESTED READING

Beilke, S., & Elshout, A.J. eds (1983). *Acid Deposition*. Reidel, Boston, Massachusetts. x + 235 pp., illus. (Distributor: Kluwer Boston, Hingham, Massachusetts) From a workshop in Berlin, September 1982.

Chevone, B.I., & Linzon, S.N. (1988). Tree decline in North America. *Environ. Pollut.*, **50**, 87–100.

Mohnen, V.A. (1986). The challenge of acid rain. *Sci. Amer.*, **255**, 30–37.

Papke, H.E., Krahl-Urban, B., Peters, K., & Schimansky, Chr. (1986). *Waldschaden*. Bundesministeriums für Forschung und Technologie, and the US Environmental Protection Agency.

Reuss, J.O., & Johnson, D.W. (1986). *Acid Deposition and the Acidification of Soils and Waters*. Springer-Verlag, New York. viii + 119 pp., illus.

Wellburn, A. (1988). *Air Pollution and Acid Rain*. Longman Scientific & Technical, Essex. xiii + 274 pp., illus.

CHAPTER 13

Ecosystem Responses

THE ECOSYSTEM

An ecosystem consists of all the organisms present in a particular area, together with their physical environment. Thus, when we talk about ecosystem responses, we are technically dealing with every living organism in a given environment. Ecological systems are energy-driven complexes whose components have evolved together over long periods of time. This close interdependency would appear to make ecosystems especially vulnerable to disruption if even one member of the system were affected. Most air pollutant-related 'ecosystem' studies involve relatively few plant species and rarely any animals or fungi. Although air pollutant effects on the dominant plants generally studied may have secondary effects on other members of the ecosystem, these have rarely been studied except for the interactions of pollutants with disease and insects. Consequently, the ecosystem responses discussed in this chapter are not as inclusive as might be inferred by the title. At best they are plant community studies, or studies that simply deal with the sensitivity of native plants.

Despite the limitations of 'ecosystem' studies, they are still basically directed towards such fundamental questions as: (1) will air pollutants ultimately play a major role in plant succession in the same way as plant diseases, insects or other environmental stresses? (2) how do pollutants affect species richness and diversity?

MECHANISMS OF AIR POLLUTANT INFLUENCE

The environment has always selected which plants will perish and which will survive. Selective pressures of climate, soil and disease are well known. The impact of one more pathogen, air pollution, is less well established but has been recognized as a potential selective force for many years (Treshow, 1968; Guderian, 1977; Materna, 1984). Disturbances caused by air pollution,

162

through stress on the individual sensitive plants, may upset the balance of the ecosystem and lead to dominance of the more tolerant species.

Primary effects

Ecosystem disruptions may be primary or secondary in nature. The primary effects are the direct effects of air pollutants on plants. The outright killing of such plants eliminates them from the system. Suppression of their vigor and growth reduces their competitive ability and has the potential for their ultimate loss from the system. The nature of the new system, or disclimax, depends on the character of the more tolerant species replacing the sensitive ones. While any effect could be considered negative, this is clearly a matter of opinion and depends on the nature of the new ecosystem. In an area of Idaho where fluorides killed several hundred acres of Douglas fir trees, grasses thrived, to the delight of local sheep ranchers.

The ultimate population of species is determined by the balance of reproduction with mortality. Such data are difficult to obtain and largely lacking, even for annuals (Harper, 1977). Reproductive and mortality data for perennial plants, most notably conifers, are far more difficult to obtain. Almost nothing is known about the relationships of air pollutants to longevity and related parameters of such species.

Rather, emphasis has been placed on the direct effects air pollutants have on the most prominent species. This has involved forests almost exclusively. Conifer forests, historically, have been severely damaged around nearly every copper smelter or aluminum reduction plant in Europe and North America (Miller & McBride, 1975). There has been little to no quantitative research to learn possible subsequent effects on the rest of the plant community or ecosystem.

Forest decline may be more chronic. Visible symptoms are absent or less apparent, and death takes place over a longer period of time. In some of these cases the affected plants may be more tolerant of a given pollutant. It is then that the pollutant may be acting indirectly and the effects are considered to be secondary.

Secondary effects

More subtle, indirect, effects may take place in a number of ways. One of the most critical of these involves changes in biogeochemical cycles. The addition of sulfur from sulfur dioxide to some systems has been taking place for many decades. Additions of nitrogen from oxides of nitrogen and ammonia are of more recent origin, but may have long-lasting consequences. Related consequences involve both the direct absorption of pollutants in the leaves and indirect absorption through the soil and roots. Uptake of oxides of sulfur and nitrogen can be significant. In the case of forest decline in Europe, one postulate of the cause suggests that the added nitrogen stimulates growth, rendering the trees more sensitive to winter injury or other stresses.

Another potential secondary effect of pollutants such as sulfur dioxide is the acidification of the soil. As a rule, soils are well-buffered systems, but this capacity depends on the soil characteristics, most notably the calcium and phosphorus composition as well as organic matter content.

Even if acidification were not significant in itself, a secondary role should be considered. This is the change in solubility of chemicals caused by acidification. This can result in mobilization of ions that become toxic in higher concentrations, such as aluminum, or the increased solubility and leaching of essential elements, such as magnesium (Rehfuess, 1981). Leaching of essential nutrients from the acidified soil disrupts the ratios of critical elements as well as causing deficiencies. Concerns of this potential impact have been limited to the dominant trees, mostly conifers, and apparently there are no studies involving whole plant communities.

Secondary effects also can be produced by altering host–parasite relations. Although this subject has been well studied on forest trees and agricultural crops, essentially no research has involved ecosystems. Mycorrhizal interactions, together with soil microfauna, must also be considered in this regard, but again facts are wanting. Along this line, changes in the rate of humus layer decomposition, and humus dynamics in general, may be important.

ECOLOGICAL RESPONSES

Desert ecosystems

At first glance deserts might seem to be the least likely candidates for air pollution impact. Not necessarily! Two factors are critical to plant response: presence of a pollutant source and sensitivity and prevalence of key species.

Pollution studies of desert ecosystems are limited to the US southwest. The region has rich coal deposits that have given rise to a number of coal-fired power-generating plants. Sulfur dioxide pollution poses one potential problem. Photochemical pollution from nearby urban areas poses another. In both cases, studies mostly have involved dominant species rather than effects on interactions in plant communities.

Plans for coal-fired power plants in the southwestern desert areas stimulated interest in the response of the native xeric vegetation to sulfur dioxide. In one study (Hill *et al.*, 1974), sulfur dioxide, nitrogen dioxide, and combinations of the gases, were introduced into chambers placed over native stands that included 87 different plant species. Concentrations of sulfur dioxide from 0.5 to 11 ppm (1311–28.84 μg/m^3) and nitrogen dioxide from 0.1 to 5.0 ppm (188–9408 $\mu g/m^3$) were studied. Visible symptoms of injury were sought following 2-hour fumigations. Few species were injured below 2.0 ppm sulfur dioxide. Only Indian rice grass (*Oryzopsis hymenoides*) and globe mallow (*Sphaeralcea malvacearum*) were injured below 1.0 ppm sulfur dioxide. This degree of sensitivity occurred only when plants were irrigated prior to fumigation to simulate an exceptionally wet year. The dominant Utah juniper showed injury only at 10 ppm sulfur dioxide, and pinyon pine at 6 ppm. Nitrogen dioxide was not injurious except in combination with sulfur dioxide; and then the effects were not synergistic. Many plants were extremely tolerant. Galleta grass (*Hilaria jamesii*), an important desert grass species, was exposed successively to 6 ppm and 10 ppm sulfur dioxide without any injury developing. Even irrigating the plants before fumigation to enhance sensitivity did not foster injury.

Historically, however, sulfur dioxide has produced some devastating effects even to deserts, but at far greater concentrations. Near copper smelters, annual and herbaceous perennial plants, grasses, cacti and some shrubs are almost absent (Wood & Nash, 1976). Not only have the plant density and cover been altered, but the species diversity as well. The direct impact of sulfur dioxide has been blamed, but there may be more complex interactions. Since many desert species are relatively tolerant to sulfur dioxide, the indirect effect on lowering the pH has been considered. In one study (Dawson & Nash, 1980), the pH of the top 2 cm of soil 2 km from a smelter was 4.0; 4 km away the pH was 4.8 and at 6 km it was 6.5. The background pH was 7.8–8.1. The disappearance of shallow-rooted annuals near the smelter well may have been due more to the acidified soil than to sulfur dioxide. Elevated concentrations of heavy metals, including copper, also may have been involved inasmuch as the extremely copper-tolerant scrub oak (*Quercus turbinella*) remained prevalent close to the smelter.

Plants of the Mojave Desert nearest the Los Angeles Basin were studied by scientists from the University of California at Riverside (Thompson *et al.*, 1980). Ten species of plants native to the Mojave Desert were exposed to sulfur dioxide at concentrations ranging from 0.22 ppm to 2.0 ppm, nitrogen dioxide at 0.11–1.0 ppm and combinations of these. Perennial plants were fumigated for up to 32 weeks and annuals for up to 17 weeks. Plants were well watered and fertilized to encourage symptom expressions. The higher concentrations of 2.0 ppm sulfur dioxide or 1.0 ppm nitrogen dioxide caused extensive leaf injury and markedly reduced growth of the dominant *Larrea divaricota* Cav. (cresosote bush), *Chilopsis linearis* Cav. (desert willow) and *Ambrosia dumosa* (Gray) Payne (Burro weed). Combinations of the pollutants had additive effects but there was no synergism. Lower concentrations stimulated growth of *Encelia farinosa* (brittlebush) and *Erodium cicutarium* (stork's bill). *Atriplex canescens* (saltbush), a common shrub throughout the southwest, was found to be highly resistant to sulfur dioxide.

Subsequent studies reflected considerable variation in sensitivity both to ozone and sulfur dioxide (Thompson *et al.*, 1984). Three species, *Camissonia claviformis*, *C. hirtella* and *Cryptantha nevadensis*, exhibited leaf injury when exposed to 100 ppb (196 μg/m³) ozone or 200 ppb (524 μg/m³) sulfur dioxide for 23–40 hours. Most species were injured following exposures above 200–300 ppb ozone or 500–1500 ppb sulfur dioxide, while six of the 37 species studied were uninjured even at 300 ppb ozone or 1500 ppb sulfur dioxide.

One of the greatest concerns involves plants growing in national parks and national monuments. Some of the more sensitive species are at risk in such parks located near large urban areas or in proximity to coal-fired power plants (Temple, 1988).

The greatest threat may be to lichens. In certain desert environments, such as Canyonlands National Park, terricolous lichens known as 'cryptogamic soil' comprise a major part of the cover that facilitates water penetration and prevents excessive run-off of the heavy rain showers that characterize that environment. Damage to these potentially sensitive lichens could result in serious erosion. Also, cryptogamic soil lichens have a critical role in nitrogen fixation in the ecosystem because their algal symbionts are mostly blue-green algae. But not all desert lichens fix nitrogen.

Heath and grassland ecosystems

Grasslands are critical to livestock production and comprise a major portion of lands in the western USA. Most are remote from ozone sources, but a few are subject to local pollution from power plants or industry. The possible impact of sulfur dioxide on grasslands was studied in Montana. Sulfur dioxide was released over plots in such a way as to expose different plots to monthly median concentrations ranging from zero to 70 ppb (185 μg/m³) (Preston, 1979). Field observations over a four-year period revealed no visible injury symptoms (Heitschmidt *et al.*, 1978). However, there was a build-up of sulfur in the fumigated plants, and some biochemical changes were noted in the dominant range plant, *Agropyron smithii* (western wheatgrass) (Lauenroth & Heasley, 1980). The change included decreases in chlorophyll content at the high sulfur dioxide levels. However, there was no change in net primary production. The single most significant alteration was a decrease in the carbon stores in the rhizomes of western wheatgrass (Lauenroth & Heasley, 1980).

A striking change in plant community structure has appeared in the heathland of the Netherlands in recent years (Heil & Diemont, 1983). Most notably, *Molinia caerulea* and *Deschampsia flexuosa* expand strongly at the expenses of *Calluna vulgaris* and other heathland species. Where there is competition between the heather species such as *Erica tetralix* and grassland species, the grasses benefit from the higher nitrogen levels (Roelofs *et al.*,

1987). Differences in nitrogen levels were striking in the grass-dominated heathlands. Ammonia levels were 10–20 times higher than the nitrate levels. A major part of this was attributed to atmospheric deposition. Nitrogen deposition increased from a few kilograms per hectare per year to between 20 and 60 kg per hectare per year, most of which is ammonium sulfate.

Woodlands and chaparral

Studies of woodlands in the western USA were occasioned by the construction of power-generating plants in such environments. The 'woodland' ecosystem in this case was pinyon–juniper (*Pinus edulis* and *Juniperus osteosperma*), forests that occupied areas expecting approximately 40–50 cm annual precipitation.

Community dynamics were studied at increased distances from a coal-fired power-generating plant for six years before and after it went on-line (Treshow & Allan, 1979). The thrust of the study was to determine if any changes occurred in the frequency or percentage cover of the different plant species. Whether or not visible symptoms of sulfur dioxide injury occurred was of secondary concern. However, no such injury appeared on even the sensitive Indian rice grass (*Oryzopsis hymenoides*), and there were no changes in community structure. This and related studies revealed that the natural annual variations in community structure, especially among annual plants, was so great that it would be difficult to detect changes caused by a pollutant except under extreme conditions. The baseline conditions themselves are highly variable (Treshow & Allan, 1985). The forb cover was unique each year and exhibited the greatest annual fluctuation. The grass cover was less variable; nevertheless, data had to be collected over a four-year period to provide a baseline wherein variation fell within an acceptable confidence level. Shrub and tree covers were relatively stable and less sensitive to change, whether from natural inputs or perturbation.

Chaparral vegetation is prominent in southern California, where ozone pollution is especially intense. Seedling chaparral plants exposed to ozone were highly variable in sensitivity. Most were considered intermediate. A potential for changes in composition and density of stands was recognized, but was not confirmed (Stolte, 1982). Since chaparral is subject to frequent fires, the post-fire seedling establishment of perennial dominants is critical to the system and most subject to disruption by ozone or other pollutant stress.

Forest ecosystems

Forests are dying! Or are they? The question has received intensive scrutiny. Forests are damaged and can be killed by any number of pathogens. This is nothing new. But attention recently has been focused on the possible association of tree deaths with acid deposition or gaseous air pollution. Whatever the causal agent, once the dominant trees are killed, demise or replacement of the understory may be reasonably expected to follow (Woodwell, 1970). Replacement of the understory by more tolerant species may have adverse effects. Water runoff, especially, may be altered. Reductions in the leaf density, or tree cover, initially will decrease interception of precipitation. The new understory subsequently may absorb more or less water than the old, with the effect of causing an imbalance in runoff, especially in protecting snow reserves in the spring and slowing the melting. Further erosion of soils under the damaged forests may result (Materna, 1984).

Despite the concern given to the dominant forest species, both around point sources and over broad regions, less attention has been given to the total ecosystem, especially at early stages of decline of understory plants. An early premise held that the overstory would be the first affected, and shrub and forb decline would subsequently follow (Woodwell, 1970). This work was conducted with chronic gamma-radiation in an oak–pine forest. Do the results also hold with other pollutants and forest types? Certainly the taller trees are subjected to the greatest amounts of pollution since they are the first to catch any atmospheric contaminants and serve as a filter for other plants. Yet, if any understory species are sufficiently sensitive to a given pollutant, they may be far more readily injured than the overstory, even at much lower concentrations.

Such was the case near Georgetown, Idaho, where the fluoride-sensitive Oregon grape (*Mahonia repens*) showed injury before any visible symptoms appeared on the dominant Douglas fir trees (Anderson, 1966). This was one of the few studies that dealt with the overall plant community responses. Along with Oregon grape, Douglas fir seedlings and various lichens and mosses, all known to be sensitive to fluoride, occurred far less frequently in areas that had been exposed to the higher fluoride concentrations. More tolerant grasses, such as pine grass (*Calamagrostis rubescens*) and forbs, especially annuals, became more plentiful in higher fluoride areas. The increasers also included *Geranium fremontii*, *Arnica cordifolia*, *Hydrophyllum capitatum* and *Osmorrhiza chilensis*. Although community structure changes were most striking where the Douglas fir trees had been killed, they were detected in stands that had no other visible symptoms. Thirteen years after the phosphate plant closed down, Oregon grape remained virtually absent, and the other changes were still apparent. Except for dense mats of moss, tolerant forbs and some choke-cherry, recovery is still barely noticeable (Treshow, 1980) (Figure 13.1).

Of the deciduous forest types, only aspen has received much attention (Treshow & Stewart, 1973). Aspen (*Populus tremuloides*) itself was the most

Figure 13.1 General absence of recovery of Douglas fir forest 13 years after cessation
of fluoride air pollution in Georgetown Canyon, Idaho.

sensitive of 40 species studied for ozone sensitivity in this community. A
single 2-hour exposure to 150 ppb (295 μg/m^3) ozone caused striking
symptoms on 30% of the foliage. This sensitivity has the potential to
predispose the trees to secondary pathogens. Loss of the aspen could also
disrupt normal succession by such shade-requiring species as white fir.

Eastern deciduous forests were studied along a pollution gradient extending
along the Ohio River Valley (McClenahen, 1978). Pollutants included
chloride, sulfur dioxide and fluoride. Species richness (number of different
species), evenness (dominance index) and the Shannon diversity index were
typically reduced within the overstory, subcanopy and herb strata near the
industrial sources of air pollutants. The importance of sugar maple (*Acer
saccharum*) was greatly reduced in all strata with increasing pollutant dose.
Yellow buckeye (*Aesculus octandra*) was most tolerant. In the shrub layer,
abundance of spice bush (*Lindera benzoin*) increased with pollutant exposure.

The conifer forests of southern California have been under intensive
oxidant stress since the 1950s. They have been most critically studied.
Ponderosa pine is the principal of five major species of this mixed conifer-
type forest. Other species include sugar pine (*Pinus lambertiana*), white fir
(*Abies concolor*), incense cedar (*Libocedrus decurrens*) and California black

oak (*Quercus kelloggii*). A 1969 aerial survey revealed that some 1.3 million ponderosa or Jeffrey pine trees (*P. jeffreyi*) on over 100 000 acres (about 40 500 hectares) were stressed to some degree. Ten per cent of the trees had died between 1967 and 1969 (Miller, 1973). White fir trees suffered light to severe damage, and the other three dominant species generally were only slightly damaged from chronic oxidant exposure. Between 1968 and 1972 mean daily maximum ozone concentrations were 200 ppb (393 $\mu g/m^3$) and generally exceeded 100 ppb 9–10 hours daily. Although it was postulated that the most sensitive plant species already may have been killed out from the community, no data regarding the understory are provided. Following loss of ponderosa and Jeffrey pines, sugar pine and incense cedar may assume greater importance. But they are not well adapted to the drier sites. More tolerant chaparral species gradually come to occupy such sites.

The influence of air pollutants on a plant community well may be related to the successional status of the forest. Some work suggests that the early successional plant species are most sensitive (Harkov & Brennan, 1979). Late successional species tend to be more tolerant. The low net production of mature or climax communities may reduce the potential impact of pollutants on photosynthesis. The slower nutrient cycling also may make the mature system less sensitive to perturbation. But there are many exceptions to this concept. Taken individually, certain species are simply highly sensitive to pollutant stress regardless of their position in the plant community.

There is no question that air pollutants can alter the composition of forests when the damage is sufficiently severe. Damage from sulfur dioxide may be subtle and occur over a period of years. But it can be blatant, as where trees are killed within just a few years. Fluoride can act similarly, and the historical record is clear (Miller & McBride, 1975). Relatively high concentrations of sulfur dioxide in the Ruhr area of Europe and the industrial Peninnes of England have influenced the distribution of Scotch pine (*Pinus sylvestris* L.). The loss of silver fir in central Europe is also well documented, although the specific cause is not always the same (Materna, 1984). In Czechoslovakia the dieback of fir begins when the mean, long-lasting concentration of sulfur dioxide exceeds 20 $\mu g/m^3$ (about 8 ppb). Perhaps just as important, exposure to such concentrations increases the susceptibility of the trees to winter stress. The greater amount of forest decline found at higher elevations appears to implicate this relationship. Concentrations of about 50–70 $\mu g/m^3$ (19–27 ppb), as long-term averages, also affect Norway spruce, leading to replacement by birch trees. Silver birch (*Betula verrucosa*) and mountain ash (*Sorbus aucuparia*) have become important in some areas comprising the new plant community. Little has been described regarding changes in the understory species.

FINAL ANALYSIS

Effects on ecosystems from air pollutant stresses can be determined only after we know the normal, constantly changing dynamics of the ecosystem. Ecosystems are characterized by structure and function. Structure can be described in terms of trophic levels and ecological pyramids of population and biomass. It involves concepts of species diversity and stability. Diversity considers richness, or numbers of species, and distribution and abundance. Function relates to productivity and biogeochemical cycles.

Details of these parameters are not well documented prior to the occurrence of air pollution. Thus the baseline data for making post-pollutant evaluations are lacking. Data available concern largely individual species, not the broad interactions involved in ecosystem analysis. Nevertheless, some tentative generalizations can be drawn (Bordeau & Treshow, 1978):

(1) Acute exposures sufficient to kill dominant species, such as has occurred around various point sources, tend to bring succession back to a previous stage. This is generally accompanied by a decrease in species richness.
(2) Chronic exposures, as exemplified by regional pollution, at locations more remote from point sources, bring about gradual modifications of ecosystem structure to adapt it to the new prevailing conditions. Species or ecotypes become replaced by more tolerant ones better able to compete in the new environment.

SUGGESTED READING

Acid Rain Foundation, St Paul, Minnesota (1985). *Air Pollutant Effects on Forest Ecosystems*. From a symposium at St Paul, Minnesota, May 1985, xviii + 439 pp., illus.

Brandt, C.J., ed. (1987). *Acidic Precipitation: Formation and Impact on Terrestrial Ecosystems*. Verein Deutscher Ingenieure, Dusseldorf, FRG, (1987). (Distributor: C.J. Brandt, 622 NW Linden Ave., Corvallis, Oregon 97330). Translated from the German edition (Dusseldorf, 1983). x + 281 pp., illus.

Draggan, S., Cohrssen, J.J., & Morrison, R.E., eds (1987). *Preserving Ecological Systems: The Agenda for Long-term Research and Development*. Proeges, Westport, Connecticut (Based on a meeting in Washington, DC, 1984) xxxvi + 192 pp.

Legge, A.H., & Krupa, S.V., eds (1986). *Air Pollutants and Their Effects on the Terrestrial Ecosystem*. Wiley–Interscience, New York. Advances in Environmental Science and Technology Series, Vol. 18. xx + 662 pp., illus.

Mathy, P., ed. (1987). *Air Pollution and Ecosystems*. Reidel (Klawer Acad. Publ. Group) 981 pp. (Papers from a meeting in Grenoble, May 1987).

Sheehan, P.J., *et al.*, eds (1984). *Effects of Pollutants at the Ecosystem Level*. Published for the Scientific Committee on Problems of the Environment (SCOPE) of the ICSU by Wiley, New York. SCOPE 22. xvi + 443 pp., illus.

CHAPTER 14

Natural Atmospheric Processes

THE NATURAL ATMOSPHERE

The natural atmosphere is seldom thought of as harmful or damaging; nevertheless, within it exist great threats to plant health and, indeed, to all life as we know it. In its present state the atmosphere is only beginning to exhibit its potential for visiting serious adverse effects upon the earth's living systems. Given the continuance of certain trends currently being imposed by human activity, the consequences of changes in some of the natural components of the atmosphere could well be more severe than the effects of all the existing industrial and domestic air pollutants combined. Thus, it is important to understand what we think of as the 'natural' atmosphere and consider our effects upon it (Clark, 1982).

The earth's atmosphere has in the past largely been taken for granted. Its composition, as usually presented, is 78% nitrogen, 21% oxygen, and 1% argon and 'other gases'. This seems simple and straightforward, but there are many complex, dynamic and significant species among those 'other gases' (see Table 14.1, which lists some additional gases in the air). These numbers refer to dry air, from which the highly variable water content has been excluded. Water vapor is, of course, an important part of the atmosphere, ranging from near zero to about 7%, depending on the temperature and other factors, but it is usually not included in the recipe for the natural composition of air.

This atmosphere, which blankets the earth to a depth of several hundred kilometers, comprises a total mass of 5.1×10^{21} g ($= 2.6 \times 10^{15}$ metric tons). At sea level its pressure is equal to that exerted by a column of mercury 760 mm (29.92 inches) high. This pressure, which is called 'one atmosphere', is equal to 1.033 kg/cm^2 (14.7 pounds per square inch).

The atmosphere is conveniently divided into layers, each of which is

172

Table 14.1. Gaseous composition of the atmosphere[1]

	Volume per cent in dry air	ppmv	ppbv[2]
Nitrogen (N_2)	78.09	780 900	
Oxygen (O_2)	20.94	209 400	
Argon (Ar)	0.93	9 300	
Carbon dioxide (CO_2)	0.034 5	345	
Neon (Ne)	0.001 8	18	
Helium (He)	0.000 52	5.2	
Methane (CH_4)	0.000 17	1.7	1700
Krypton (Kr)	0.000 10	1.0	1000
Hydrogen (H_2)	0.000 05	0.5	500
CFC-12 (CCl_2F_2)	0.000 038	0.38	380
Nitrous oxide (N_2O)	0.000 030 7	0.307	307
CFC-11 (CCl_3F)	0.000 022	0.22	220
CH_3CCl_3	0.000 013 3	0.133	133
Carbon tetrachloride (CCl_4)	0.000 012 3	0.123	123
Xenon (Xe)	0.000 009	0.09	90
Tropospheric ozone (O_3)	0.000 003 5[3]	0.035	35
Other organic vapors	0.000 002	0.02	20
Total	99.997	999 972.708	

[1] Data after Ramanathan (1988) and Stern *et al.* (1984). This table is for unpolluted dry air. For wet basis composition at 1 atm. pressure (= 760 mmHg), the relative proportions of each atmospheric component may be adjusted by using a water vapor value in the range from near zero up to about 7% and proportioning the remaining fractions accordingly by keeping the total at 100%. The vapor pressure of water at 0, 25 and 40 °C = 4.6, 23.8 and 55.3 mmHg, respectively. At 25 °C, for example, a representative value of 3.13% water is obtained by 23.8/760 × 100 = 3.13%. At 0 °C, water = 0.6%; and at 40 °C, water = 7.3%. Water content in the atmosphere is, of course, highly variable at any temperature.
[2] ppb = parts per billion; 'billion' means 10^9 or 'thousand million'.
[3] Tropospheric (under 12 km altitude) ozone is variable, depending on precursors, scavengers and insolation, ranging from about 0.010 to 0.100 ppm. We have measured day and night background levels of ozone in remote areas of southern Utah at an average of 0.035 ppm, which is the value given in the table.

defined more by the physical and chemical processes that occur there than by any strict altitudes. Near the ground is the 'troposphere', which extends upward about 12 km. All 'weather' takes place within the troposphere. This is the region of surface winds, clouds, rain, thunderstorms, frontal systems, hurricanes, tornadoes and all the familiar phenomena that appear on the daily weather report. Above the troposphere and extending to about 50 km is the 'stratosphere', a region of rarified air that is characterized especially by a layer of ozone (O_3), produced photochemically by the high-energy radiation present at these elevations. This ozone layer, which is most developed at an altitude of about 26 km, serves to shield the earth from damaging ultraviolet radiation. Without this stratospheric ozone, excess ultraviolet radiation would reach the earth's surface to penetrate and disrupt living cells. Ultraviolet radiation can disrupt many important biological molecules, including DNA, chlorophyll, other plant pigments and proteins. Not only would skin cancer increase, but in aquatic habitats the biomass of green algae, diatoms and other phytoplankton would be reduced.

Above the stratosphere, and extending outward into space for hundreds, perhaps thousands, of kilometers, are regions called the mesosphere, thermosphere and exosphere. Here, the intense radiation from the sun and from cosmic rays disrupts atoms of hydrogen, helium and oxygen, as well as others, to create ions—electrically charged entities that move along the earth's magnetic lines towards the poles, where they create spectacular auroras.

The oceans of the earth cover some 70% of the globe, and their combined mass amounts to 1.4×10^{24} grams ($= 7.0 \times 10^{17}$ metric tons). They are linked to the atmosphere through their intimate mutual contact, and by the turbulence that exists at the ocean surface. Across the interface between ocean and air pass water molecules, carbon dioxide, oxygen, oxides of nitrogen, and the many gaseous and particulate pollutants that are injected into the air by the activities of humans. These exchanges move in both directions; water evaporates from the ocean surface and airborne gases dissolve into it. Rain falls upon it, carrying both dissolved atmospheric gases and particulate matter (Betzer *et al.*, 1984).

The atmosphere and the ocean together constitute a vast reservoir of chemicals that, depending upon the energy available and the chemical equilibria that exist, act to maintain a kind of balanced condition over the whole earth (Goldman & Dennett, 1983). For example, about 0.15% of the weight of sea water is carbon dioxide, mostly in the form of the hydrogen carbonate ion (HCO_3^-). This is a little more than one thousand times the amount of carbon dioxide found in the atmosphere. In addition to dissolved carbon dioxide, the sea contains very large quantities of carbon dioxide bound in carbonate rocks such as limestone. A fundamental or long-term shift in the climatic conditions of the earth could very well alter the ocean–air equilibrium of carbon dioxide, leading to large releases of carbon dioxide from the oceans and even from oceanic rocks that would be subject to dissolving processes.

The ocean–air system is dynamic, to be sure; it is not uniform or static either with respect to carbon dioxide or other chemical species. The water vapor content in the atmosphere, for example, is highly variable, depending on local air and surface temperatures, air movement and atmospheric pressure. Ninety per cent of all atmospheric water vapor is concentrated below 6000 m (about 19 700 feet) altitude, which means that the earth's mountain ranges can be important barriers to its movement. Atmospheric pressure is itself highly variable. The surface temperature is affected in turn by the water vapor itself. Clouds form and reflect sunlight energy away from the earth. Yet at the same time clouds serve to trap and re-radiate heat energy back towards the earth. Like the ocean, the atmosphere is not simple, either in its dynamics or its composition.

Radiation from the sun is undoubtedly the most important factor in the dynamics of the atmosphere. It is the source of the power that drives both atmospheric and oceanic circulation. It is the source of energy that drives the hurricane and creates storm clouds. The sun's annual excursions to the north and south of the equator, due to the earth's tilted axis, which carry the direct impact of solar radiation with it, alternately warm the northern and southern hemispheres. Ice and snow are melted, water flows to the sea, sea and land are warmed, and water vapor is evaporated into the atmosphere. The vegetation, both terrestrial and aquatic, grows and thrives. Photosynthesis proceeds at its maximum rate. In the cold hemisphere during these cycles, days shorten, plants enter dormancy, the land and sea cool, clouds form, snow falls and accumulates, ice freezes.

Incoming solar radiation is not the only kind of radiation important to the earth. Outgoing radiation from the earth is equally critical. If there were no outgoing radiation, the earth must inevitably continue to increase in temperature, which it does not do. An equilibrium of sorts exists; it is a variable equilibrium to be sure, with temperature excursions lasting from but moments to thousands of years. But on the whole the temperature of the earth remains essentially constant. Evidence exists that the oceans of the earth have never either frozen solid nor boiled. Thus, between these two extremes there exists a kind of heat budget balance that has existed throughout the 4.5 thousand million year-long history of the earth.

Incoming radiation is balanced by outgoing radiation on the whole, with the temporary imbalances exhibiting themselves in the form of ice-ages or tropic-like interglacial periods. Even the age-long tropic-like humid eras such as those that gave us our coal, oil and natural gas deposits are but eras of more or less balanced energy budgets for the earth. The coal forests of the Pennsylvanian, Mississippian and other geological ages existed during times when photosynthetic fixation of carbon exceeded decay, fire and other processes that would have released carbon—as carbon dioxide—back to the atmosphere. Photosynthesis uses sunlight energy to convert atmospheric carbon dioxide to sugars, from which starch, proteins and other plant substances, including wood, which is mostly cellulose, lignin and other carbon-based compounds, are produced.

Energy is thus an important aspect of the environment of the earth, not only as received from the sun and re-radiated to space, but also as exchanged between earth and the atmosphere, the atmosphere and the ocean, and between living things and the environment. In addition to the atmosphere, the ocean, and the energy provided by the sun, the land area of the earth provides the fourth element of the earth environment. Source of the minerals necessary to life, habitat of terrestrial animals and plants, including humans and forests, and seedbed for food crops, the land is also important in the grand cycles of nature. When this consideration is added to our understanding of the dynamic systems of ocean, atmosphere and solar radiation, we find that our existence is dependent upon the original four elements of the ancient Greeks: earth, air, fire (energy) and water. Perhaps they were not so far from the truth after all—the dynamics of our planet are encompassed in the balances that exist between the earth, the atmosphere, the oceans and the energy budget of the whole.

ENERGY BALANCE OF THE EARTH

This review of the natural atmosphere leads logically to the energy balance of the earth as it is mediated by the atmosphere. It leads also to a consideration of the consequences of any imbalances that might be created by excesses or deficiencies in any of the physical or chemical components of the atmosphere.

Heat, ultimately derived from solar radiation, can be transferred to the atmosphere in four different ways. One is by conduction, the process by which heat travels along an iron rod placed in a fire, by the direct motion of molecules. Temperature is nothing more than the motion of molecules; the faster they move, the higher the temperature. These lively molecules strike the slower ones further along the rod, knocking them into more vigorous motion as well, which we sense as rising temperature along the rod.

The second form of heat transfer is similar, but it involves actual movement of some of the mass of the substance being heated, not just molecular motion as in the iron rod. This is convection; water heated in a beaker is a good example. It is the same kind of conduction as in the iron rod, but a stream of water rises from the bottom of the beaker, carrying heat with it. It is a process common to both liquids and gases. It is almost always present in the atmosphere where air is warmed near the ground, causing it to expand and rise. New, cooler air rushes in around the rising air column or air bubble—which can be hundreds of meters across—to start the process over again. The rising air, sometimes called a 'thermal', can be a powerful upward source of movement, not only for the air mass itself but for dust, pollen, weeds, papers, plastic and other materials near the ground. Some of these rising air currents are real 'trash-movers'; dust devils are a good example. These are rising air currents that twist like small tornadoes, carrying aloft almost anything in their path. These are generated by solar heating of the ground surface followed by convection, the mass movement of the atmosphere as the heated air rises. Night-time air is often found in the morning to have layered near the ground in cool strata containing the settled dust and captured emissions of automobiles and home heating. It is the convective processes of the morning's sunlit heating that breaks up these inversions and disperses them, often by mid-morning.

A third way by which heat is transferred to the atmosphere is by means of evaporation. Water requires and absorbs about 580 calories of heat from the earth's surface (and from incoming solar radiation) for each gram of water converted to water vapor, which then enters the atmosphere. The same is true of water being transpired (evaporated) from the surface of a leaf—or, more accurately, from the surfaces of the cells within a leaf. Whenever water undergoes the phase change from liquid to vapor, 539 calories of heat energy per gram of water are absorbed by the molecules of water. The difference between 539 and 580 represents heat taken up by the water to raise some of its molecules from ambient temperature to the temperature at which the phase change to gas can occur. All of this heat enters the atmosphere with the water vapor.

Water vapor rises in air, as does air that contains some water vapor. It is lighter than air, just as helium or hydrogen is, but of course not by so much of a difference. Air is about 99% nitrogen and oxygen—78% nitrogen and 21% oxygen. The remaining 1% contains all the other permanent, natural components of the atmosphere that are normally included: argon, carbon dioxide, hydrogen, neon, helium, krypton, ozone and xenon, listed in descending order of abundance. Nitrogen atoms 'weigh' 14 units each, i.e. the atomic weight of nitrogen is 14. Since the normal form of atmospheric nitrogen is N_2, or dinitrogen, the molecular weight of atmospheric nitrogen is 28. The atomic weight of oxygen is 16. Atmospheric oxygen is also a diatomic molecule with molecular weight 32. The molecular weight of air is thus about $0.78 \times 28 + 0.21 \times 32 = 28.56$. Water vapor, which is composed of molecules of water, H_2O, has a molecular weight of $2 + 16 = 18$, which is only about 63% of the molecular weight of air. Accordingly, water vapor, or air containing water vapor, naturally tends to rise in the atmosphere, carrying with it the latent heat of vaporization the water acquired in the phase change from liquid to vapor. This rising moist air, incidently, contributes to the mixing of the atmosphere by its passage.

The latent heat of vaporization is given up to the atmosphere when the water vapor condenses in the upper atmosphere to form clouds or precipitation. More heat—another 80 calories per gram of water—is given up if the cloud droplets freeze to form snow or hailstones. These phenomena account for such folklore as 'the warm before the storm' and the warm dry winds that occur in the lee of mountain ranges that intercept precipitation— winds variously called 'Chinook' or 'Santa Ana' winds. These are air masses that have given up their moisture but have retained the latent heat of vaporization from that moisture. When such air masses descend to lower elevations down the lee or rain shadow side of a mountain range, they are compressed as they descend into the lower elevations and become warm—doubly warm—first by the compression and second by the latent heat of vaporization they have acquired.

Finally, there is the process of thermal radiation, which, unlike the previous three forms of heat transfer, does not require any material carrier like the molecules of the iron rod, the water or the water vapor. Solar radiation spreading throughout space, and the radiation felt at a distance from a hot surface, are examples. Thermal radiation is electromagnetic radiation having wavelengths in the range from about 0.2 to 100 μm. This extends on both sides of the range of the visible light spectrum, which is from 0.36 to 0.76 μm (360–760 nm). Visible light is therefore also partly thermal radiation, but it only occupies a small part of the entire thermal spectrum. Wavelengths shorter than 360 nm include ultraviolet, X-rays, gamma rays and cosmic rays. Those longer than 760 nm include infrared radiation and radio waves.

The earth's atmosphere is largely transparent to visible light, including much of the thermal component it carries. Nevertheless, the atmosphere reflects, scatters and absorbs some of the solar radiation that passes through it. The atmosphere is not equally transparent to the entire electromagnetic spectrum; in particular, it is not transparent to the full bandwidth of the thermal spectrum. Even the visible spectrum is differentially transmitted. The blue of the sky is caused by light scattered by air molecules. The red of sunsets is caused by the scattering of the shorter, bluer wavelengths of light by the thicker layers of atmosphere that sunlight must penetrate at sundown. The longer, red wavelengths penetrate further through the atmosphere before being scattered than do the shorter, blue wavelengths.

ATMOSPHERIC IMBALANCE AND 'GREENHOUSE' WARMING

And now we approach the goal of this background discussion: to understand the potential impacts of imbalances in some of the natural and anthropogenic atmospheric components on plants and on all life forms.

Certain gases in the atmosphere, including water vapor, ozone, methane, carbon dioxide, and several recently manufactured chemicals (e.g. chlorofluorocarbons, creations of the last century or so) all absorb thermal radiation in various bandwidths along the thermal spectrum. And therein lies a tale, for it is thermal radiation that has to do with the 'greenhouse warming' caused by the build-up of these gases, some of which are natural and some anthropogenic. All of these gases exist in the parts per million or even parts per billion range in the less-than-1% fraction of 'other gases' that make up the composition of the atmosphere.

The well-publicized greenhouse effect is predicted to raise average earth temperatures by some 1.1–4.5 °C (Ramanathan, 1988). The consequences of this seemingly small temperature elevation are impressive, well beyond first expectations. One major impact would be the melting of the polar ice-caps, accompanied by worldwide rising sea levels, as much as 5–6 m. This would flood coastal areas, sea-coast cities, and possibly whole countries such as Bangladesh and the Netherlands—unless engineering works are initiated in time to minimize the impact (Hansen *et al.*, 1983). Still another effect of the greenhouse warming will be the accompanying displaced patterns of temperature and precipitation and, consequently, of agriculture and natural vegetation provinces. No air pollutant, however phytotoxic, widely distributed or health-threatening, compares in destructive potential to this global disaster.

This 'greenhouse warming' is not yet unambiguously detected, but many indications suggest that it has already begun (Rasmussen *et al.*, 1986). Some of these include the worldwide retreat of glaciers during this century, a perceptible rise in sea level during the last century (15 cm), and a decade (the 1980s) of widely variable weather patterns across the globe, unlike any ever before observed or inferred from the geological record. The 1980s have produced weather that has been both warmer and colder, wetter and dryer, and with record-breaking extremes in different regions of the world. Comparison of these recent weather data with earlier records has led climatologists to declare that this much variability would not occur by chance once in a thousand years. This suggests that the effect is definitely not by chance; something caused it (Robock, 1983).

Understanding the physics of what may be happening should be helpful. Sunlight falls upon a surface perpendicular to the sun's rays, at the mean earth–sun distance of 1.496 × 10^8 km, with an average intensity of about 2 calories per square centimeter per minute. This quantity of energy is called the solar constant, k. At the earth, this insolation is intercepted by a cross-sectional area equal to πR^2 where R is the earth's radius of 6371 km. Insolation at the earth is not actually constant; it varies by about ±7% as the earth revolves about the sun, for the earth's orbit is not perfectly circular but slightly elliptical. The same average amount of energy, subject to diurnal, seasonal and long-term climatic variation, is re-radiated from the earth out to space in all directions. By means of these concepts and the Stefan–Boltzmann law of black body radiation, which relates the amount of energy emitted from a surface to its absolute temperature, it is possible to determine the average temperature of the earth's surface.

The Stefan–Boltzmann law is given by $S = \sigma T^4$ in which S is the total intensity of radiation emitted over all wavelengths, T is the absolute temperature (K), and σ is the Stefan–Boltzmann constant (8.13 × 10^{-11} cal K^{-4} cm^{-2} min^{-1}). In order for the earth to remain at a long-term relatively constant average temperature, the amount of energy (S) re-radiated by the earth's spherical surface (area $4\pi R^2$) must equal the quantity of energy intercepted by its cross-sectional area (πR^2) multiplied by the solar constant, k. The relationship is $S \times 4\pi R^2 = k \times \pi R^2$. S is, of course, σT_E^4, with T_E being the temperature of the earth, which is determined as follows:

$$\sigma T_E^4 \times 4\pi R^2 = k \times \pi R^2$$

$$T_E^4 = 4k/\sigma$$

$$T_E^4 = \frac{2\,\text{cal cm}^{-2}\,\text{min}^{-2}}{4 \times 8.13 \times 10^{-11}\,\text{cal cm}^{-2}\,\text{min}^{-1}\,\text{K}^{-4}}$$

$$T_E = [6.15 \times 10^9\,\text{K}^4]^{0.25}$$

$$T_E = 280\text{K} = 7\,°\text{C}$$

The average temperature of the earth near its surface is, of course, much higher than 7 °C; it is around 15 °C, and the upper atmosphere beginning at the 11–12 km-high boundary between the troposphere and stratosphere is correspondingly colder (−50 to −80 °C). Although the overall quantity of energy leaving the earth is, and must be, equal to that received from the sun—a requirement for maintaining thermal equilibrium—the wavelengths of the incoming and outgoing radiation are very different. Solar radiation is spread over a wide range of wavelengths extending from 200 to 2200 nm, which includes the visible spectrum (360–760 nm) and peaking at the wavelength of blue-green light (470 nm). As will be seen below, the earth's 15 °C surface emits radiation at a wavelength centered at 10 μm (10 000 nm), which is at the boundary between the near infrared and the far infrared.

Wavelengths of maximum emission for any temperature can be calculated from the relationship known as Wein's displacement law, which states that the product of the absolute temperature (T) of a radiating body and the wavelength corresponding to the maximum intensity of that radiation ($\lambda_{\max}$) is constant. With T in kelvins and $\lambda_{\max}$ in micrometers, Wein's displacement law is given by:

$$T\,\lambda_{\max} = 2880\,\mu\text{m K}$$

The earth's surface, with an average temperature of 15 °C (= 288 K), thus radiates maximally at $2880/288 = 10.0$ μm. The upper atmosphere, with a temperature of −55 °C (= 218 K), radiates maximally at $2880/218 = 13.2$ μm. The land areas of the earth, its oceans and the different levels of its air masses are not, of course, always at some average temperature. The daily cycles of light and dark, the passage of the seasons, the effects of latitude that extend from well below freezing to tropical heat, and the multitude of other factors that affect temperatures on earth all operate to cause the simultaneous emission of radiation from the earth at many wavelengths at once. All of these emissions associated with the full range of land and ocean surface to stratospheric air temperatures found on earth fall within wavelengths of the infrared region of the spectrum extending from about 7 μm out to approximately 100 μm.

Wein's displacement law as given above does not give the wavelength of the *center* of the emission band associated with a given temperature, but only of the wavelength of *maximum* emission. In this form, about 25% of the total radiation is found in wavelengths shorter than λ_{max} and 75% is above λ_{max}. To find the center wavelength (λ_s), such that 50% of the total intensity lies on either side of it, the 'constant' (2880) in Wein's displacement law should be adjusted to 4100 (Hofmann, 1955). In this form the displacement law is given as $T\lambda_s = 4100$ μm K. Either form is used in different discussions of the earth's radiation to space, and the resulting energy distribution curves look quite different. For example, it was shown above that the earth's 15 °C average temperature results in a radiation wavelength of 10 μm. At this wavelength, 25% of the total intensity is below 10 μm, in the shorter wavelengths, and 75% is found in wavelengths longer than 10 μm. The wavelength for which 50% of the total intensity lies on either side, as given by the adjusted form of Wein's displacement law, is $\lambda_s = 14.2$ μm.

If the atmosphere were perfectly transparent to wavelengths of thermal radiation in the micrometer range, the earth would attain an equilibrium temperature very much colder than it is now, about -18 °C, cold enough to cover the planet with ice. But the atmosphere, including clouds, water vapor, and all natural and normal components within it, *already* exhibits a greenhouse forcing effect by absorbing some of the outgoing thermal radiation and warming the earth from -18 °C up to about 7 °C, as has already been shown. The problem faced by the world today is the *additional* warming that is being produced by several natural anthropogenic gases that are being injected into the global atmosphere by human activities. Thus we should say that we face not merely a greenhouse effect, but a *super*-greenhouse effect. Life on earth depends upon the atmosphere's natural greenhouse forcing effect, but human activities are over-intensifying it.

The unpolluted, pre-industrial atmosphere was quite transparent to infrared radiation in the spectral region from 7 to 13 μm, and about 80% of the radiation emitted from the earth's *surface* in this 'window' escaped to outer space (Ramanathan, 1988). The remaining 20% was, and is, absorbed by the atmosphere, warming it. Outside this 'window', most of the surface radiation is absorbed by the atmosphere and then re-emitted to space at colder, upper atmospheric temperatures.

Carbon dioxide absorbs strongly in the wavelengths between 13 and 17 μm, and ozone absorbs strongly between 9 and 10 μm. Water absorbs radiation in the entire infrared spectral domain. These three gases are normally thought of as 'natural' atmospheric components. Unfortunately, there are several synthetic gases that absorb strongly in the 7–13 μm 'window' region—even more strongly than does carbon dioxide in its absorption band. For example, if carbon dioxide were to double from 300 to 600 ppm, the radiative energy it would trap has been estimated (Bolin *et al.*, 1986) to be about 0.006 cal cm^{-2} min^{-1} (4 W m^{-2}). By comparison, and showing the relatively more dramatic ability of chlorofluorocarbons (CFCs) to absorb thermal energy, if CFC-11 and CFC-12 were to increase from 0 to 2 ppb (parts per billion = parts per thousand million), their thermal absorption would be about 0.001 cal cm^{-2} min^{-1} (1 W m^{-2}). These gases, at less than a thousandth of the carbon dioxide concentration, nevertheless absorb one-quarter as much thermal energy.

The chlorofluorocarbons are such effective thermal energy absorbers partly because their absorption bands are located directly in the 7–13 μm wavelength 'window' region, literally dirtying the window. Among the chlorofluorocarbons, CFC-11 and CFC-12 are the most abundant, accounting for about 82% of all CFCs produced worldwide; CFC-22 ranks third. 'Freon' is a trade-name referring to a broad class of chlorofluoromethanes. Many of the natural and synthetic gases having greenhouse forcing properties are listed in Table 14.2.

The carbon dioxide connection

Carbon dioxide is a colorless, odorless gas that is produced during combustion processes by the oxidation of carbon from organic sources. It also is produced by yeasts and other microorganisms during the anaerobic conversion of sugars into alcohol and carbon dioxide. It combines with water to form carbonic acid (H_2CO_3), a weak acid whose salts form the important carbonate minerals, including calcium carbonate (calcite, marble, limestone, coral, most sea shells and aragonite), sodium carbonate (washing soda, sal soda and soda ash) and sodium hydrogen carbonate (baking soda and bicarbonate of soda). Carbon dioxide is easily soluble in water, one liter of which at O °C dissolves 1713 ml of carbon dioxide at one atmosphere of pressure.

The atmosphere contains about 0.03% carbon dioxide by volume. Carbon dioxide has increased from about 270 ppm in about 1750 to about 345 ppm in 1984 (Detwiler & Hall, 1988). The concentration of carbon dioxide in the atmosphere is steadily increasing at a rate of 1.5 ± 0.2 ppm per year (Mooney *et al.*, 1987). It is expected to rise to some 600 ppm before the middle of the twenty-first century, and according to models (Ramanathan, 1988), this rise is expected to increase the mean global temperatures by 1.1–4.5 °C. The consequences of this runaway greenhouse effect are appalling. The list is reminiscent of the worst planetary destructions imagined by science-fiction writers. For example, the melting of the polar ice-caps that will result in raising sea levels by some 6 m, enough to inundate not only beaches and existing harbor facilities but whole coastlines with their coastal cities. Even whole countries such as Bangladesh and the Netherlands could be completely inundated (Seider & Keyes, 1983).

To the reservoir of free carbon dioxide in the air more is added by the decay of plants; by plant respiration; by animal respiration and decay; and by combustion in the burning of wood, peat, green plants and fossil fuels (coal, oil, natural gas). Some comes out of solution in the oceans of the world, and small amounts are supplied to the atmosphere as a result of volcanic activity. Carbon dioxide is removed from the atmosphere by photosynthesis of green plants and by chemical equilibrium processes with carbon dioxide dissolved in the oceans of the world. These processes have been called the carbon cycle.

Some of the carbon withdrawn from the atmosphere is not returned directly to the atmosphere but is stored in various forms. Deposits of peat, coal and oil are natural reservoirs of carbon derived from the carbon dioxide of the air. Some carbon dioxide is returned to the air by the respiration of animals and green plants, and by the combustion of wood, coal and oil. Contrary to popular belief, the total quantity of carbon dioxide released into the atmosphere by land plants is greater than that released by animals. About 20% of the carbon dioxide used in photosynthesis is returned to the air by plant respiratory processes.

In addition to the climatic effects of carbon dioxide, the rising levels of carbon dioxide in the atmosphere are expected to increase rates of photosynthesis in vegetation (Lemon, 1983). Carbon dioxide is food for plants, and it is one of the rate limiters of photosynthesis when in short supply. Increased growth response to artificially elevated carbon dioxide levels, which is called the fertilization effect, has been observed in greenhouse studies. Increased radial growth has also been observed in conifer trees growing at high altitudes (LaMarche *et al.*, 1984). The observation is consistent with observed global trends in carbon dioxide since the mid-1800s, and some (LaMarche *et al.*, 1984) believe that the increased growth is caused by rising carbon dioxide concentrations in the atmosphere.

Mitigating and reducing the consequences of carbon dioxide

'Plant trees. The tree of life is no longer a metaphor—it is now a reality.' This was the advice given by environmental activist Jeremy Rifkin when interviewed for Monitoradio by the Christian Science Monitor in October 1988. Trees remove carbon dioxide from the atmosphere. It is a tragedy that even as the burning of fossil fuels is increasing carbon dioxide at rates far exceeding the ability of natural processes to remove it, forests are also being cleared for lumber, charcoal and farming at a breath-taking rate in

Table 14.2. Some infrared-trapping trace gases in the atmosphere, with alternative names, chemical formulae, molecular weights, approximate absorption bands, and their relative absorption strengths (μm = micrometers wavelength)

Carbon tetrafluoride, refrigerant–14 or R–14, Freon–14. (CF_4, mol. wt. = 88.011)
 Absorption band: 7.8–8.3 μm Relative absorption strength = 4500

Fluoroform, trifluoromethane. (CHF_3, mol. wt. = 70.019)
 Absorption band: 8.4–9.1 μm Relative absorption strength = 3400

Hexafluoroethane, Freon–116, R–116. (CF_3CF_3, mol. wt. = 138.022)
 Absorption bands: 7.5–8.2 μm Relative absorption strength = 3300
   8.6–9.3 μm Relative absorption strength = 1100
   13.4–14.2 μm Relative absorption strength = 140

Trichlorotrifluoroethane, Freon–13, chlorofluorocarbon–13 or CFC–13, R–13.
   (CCl_2FCClF_2, mol. wt. = 187.393)
 Absorption bands: 8.2–8.6 μm Relative absorption strength = 2600
   8.7–9.3 μm Relative absorption strength = 2500
   13.8–14.4 μm Relative absorption strength = 2500
   12.5–13.1 μm Relative absorption strength = 160

Bromotrifluoromethane, Freon–13B1, R–13B1. ($CBrF_3$, mol. wt. = 148.927)
 Absorption bands: 7.9–8.7 μm Relative absorption strength = 2100
   9.1–9.9 μm Relative absorption strength = 2000

Trichlorofluoromethane, Freon–11, R–11, CFC–11. (CCl_3F, mol. wt. = 137.382)
 Absorption bands: 10.8–11.2 μm Relative absorption strength = 2100
   9.3–9.8 μm Relative absorption strength = 760

Dichlorodifluoromethane, Freon–12, CFC–12, R–12. (CCl_2F_2, mol. wt. = 120.925)
 Absorption bands: 10.8–11.3 μm Relative absorption strength = 1500
   9.1–9.6 μm Relative absorption strength = 1400
   8.3–8.8 μm Relative absorption strength = 760

Carbon tetrachloride, tetrachloromethane. (CCl_4, mol. wt. = 153.839)
 Absorption band: 12.4–13.2 μm Relative absorption strength = 1400

Methylene fluoride, difluoromethane. (CH_2F_2, mol. wt. = 52.027)
 Absorption band: 9.5–10.3 μm Relative absorption strength = 1100

Chloroform, trichloromethane. ($CHCl_3$, mol. wt. = 119.390)
 Absorption bands: 12.3–13.2 μm Relative absorption strength = 1100
   7.8–7.6 μm Relative absorption strength = 140

Sulphur dioxide. (SO_2, mol. wt. = 64.07)
 Absorption bands: 7.2–7.7 μm Relative absorption strength = 750
   8.5–9.0 μm Relative absorption strength = 100

Acetylene. C_2H_2, mol. wt. = 26.04)
 Absorption bands: 13.1–13.7 μm Relative absorption strength = 700
   <7.0–7.7 μm Relative absorption strength = 80

Chlorodifluoromethane, Freon–22, R–22, CFC–22. ($CHClF_2$, mol. wt. = 86.476)
 Absorption bands: 8.8–9.3 μm Relative absorption strength = 640
   12.0–12.5 μm Relative absorption strength = 240
   7.0–7.6 μm Relative absorption strengt = 130

Nitric acid. (HNO_3, mol. wt. = 63.016)
 Absorption band: 7.4–8.0 μm Relative absorption strength = 550

Hydrogen cyanide.		(HCN, mol. wt. = 27.027)	
Absorption band:	13.9–14.4 μm	Relative absorption strength	= 500
Ethylene.		(C$_2$H$_4$, mol. wt. = 28.05)	
Absorption band:	9.8–10.4 μm	Relative absorption strength	= 420
Methylene dichloride.		(CH$_2$Cl$_2$, mol. wt. = 84.941)	
Absorption bands:	12.7–13.6 μm	Relative absorption strength	= 340
	7.1–8.0 μm	Relative absorption strength	= 100
	13.3–12.2 μm	Relative absorption strength	= 33
Peroxyacetyl nitrate, PAN.		(CH$_3$COO$_2$NO$_2$, mol. wt. = 121.054)	
Absorption bands:	8.5–9.2 μm	Relative absorption strength	= 340
	7.2–7.7 μm	Relative absorption strength	= 310
Ozone.		(O$_3$, mol. wt. = 48.00)	
Absorption bands:	9.4–9.8 μm	Relative absorption strength	= 330
	12.4–12.8 μm	Relative absorption strength	= 320
	8.7–9.1 μm	Relative absorption strength	= 12
Methylchloroform.		(CH$_3$CCl$_3$, mol. wt. = 133.417)	
Absorption bands:	13.3–14.3 μm	Relative absorption strength	= 310
	8.8–9.8 μm	Relative absorption strength	= 150
	7.2–8.3 μm	Relative absorption strength	= 15
Hydrogen peroxide.		(H$_2$O$_2$, mol. wt. = 34.02)	
Absorption band:	7.7–8.4 μm	Relative absorption strength	= 300
Carbon dioxide.		(CO$_2$, mol. wt. = 44.01)	
Absorption band:	13.2–16+ μm	Relative absorption strength	= 240
Methane.		(CH$_4$, mol. wt. = 16.04)	
Absorption band:	7.3–7.8 μm	Relative absorption strength	= 200
Nitrous oxide.		(N$_2$O, mol. wt. = 44.02)	
Absorption band:	7.9–8.4 μm	Relative absorption strength	= 200
Ethane, R–170		(C$_2$H$_6$, mol. wt. = 30.07)	
Absorption band:	11.7–12.3 μm	Relative absorption strength	= 32

the tropics, at rates approaching 50 hectares (124 acres) a minute, worldwide (estimated from FAO data in Table 1, Detwiler & Hall, 1988).

Rifkin's advice is sound. Trees are the terrestrial equivalent of the ocean's buffering system for carbon dioxide. Just as the ocean has the ability to absorb and tie up carbon dioxide from the atmosphere, so do trees. Trees and other woody plants, more so than crop plants or herbaceous vegetation, deposit carbon in their wood in the form of cellulose, lignin and other relatively permanent substances. This wood can form a huge reservoir of carbon that would otherwise exist as atmospheric carbon dioxide. Clearing, which exposes much of the wood and all of the slash to burning and decay processes, are reconstituting this carbon as carbon dioxide and returning it to the atmosphere. The lumber obtained from the timber will itself eventually yield to burning, decay, termites, shipworms, or any of dozens of other destructive forces, all of which convert some or all of it to carbon dioxide.

Alexander L. Howard (1948) wrote in the preface to his book on commercial timber trees:

'. . . Up till now the march of civilization has everywhere proclaimed the destruction of trees over the wide surface of the globe, and the successive generations of all races have continuously failed to establish any proper measures for reafforestation. Trevelyan says: "A bird's-eye view of England in Anglo-Saxon times would have revealed a shaggy wilderness of forest trees, brush-wood, marsh and down, spreading from shore to shore." Very different is the prospect viewed by the airman to-day. "But in those days there was hope of the future, for though elm and oak might fall, men planted others—AS THEY SELDOM WILL TO-DAY." Gabrielle Plattes, writing in 1639, remarks in his book, *A Discovery of Infinite Treasure*: "Now the multitude of timber brought yearly from Norway, and other parts, doe plainly demonstrate the scarcities thereof here; also it may be conjectured what a miserable case the Kingdom will be plunged into about an age or two hence, for want of timber." About 270 years ago (i.e., in about the year 1680) John Evelyn wrote; "For I observe there is no part of husbandry, which men commonly more fail in, neglect, and have cause to repent of, than that they did not begin planting betimes, without which they can expect neither fruit, ornament, or delight from their labours." In 1853 an unnamed writer of a book, entitled *English Forests and Forest Trees*, in the introduction says: "These forests are very rapidly passing away. At present few of those that were once so famous still exist. The fine forest of Sherwood was sold in 1827; scarcely a year passes by but enclosures are made, or some forest is disafforested; and very soon not one of the forests will retain its primitive appearance."' (Howard, 1948).

There are those who claim that forests are inconsequential in the matter of global carbon dioxide balance. The US Environmental Protection Agency, for example, in reviewing a proposal on the possibility of storing significant quantitites of carbon dioxide in biomass (US Environmental Protection Agency, 1983), concluded that 'sequestering atmospheric CO_2 by trees is an extremely expensive, essentially infeasible option for controlling CO_2.' This opinion agreed with that of another cost-conscious government agency (National Research Council, 1983), which wrote, 'One conclusion is inescapable, irrespective of a hundred years' technological change, "sweeping" the atmosphere with trees can be no great part of any solution to the CO_2 problem.'

In the ocean, not only is carbon dioxide dissolved in the water itself, but it can be taken up by phytoplankton and converted into organic and inorganic carbon compounds. There is an equilibrium between carbon dioxide in the ocean and that in the atmosphere. If atmospheric carbon dioxide declines, the ocean gives up some more from its dissolved reserves, which amount to some 30 times the amount found in the atmosphere. If the ocean did not do this, the forests of the world, *as they once existed*, would be able to remove all of the atmospheric carbon dioxide in about 30–35 years.

Conventional wisdom once held that if carbon dioxide levels increased in the atmosphere, oceanic carbon dioxide would precipitate as calcium carbonate or one of the other carbonates, thus maintaining the equilibrium between oceanic and atmospheric carbon dioxide. Now that carbon dioxide is building up in the atmosphere, and greenhouse warming is already apparently being felt, no one is quite so sure any more that the ocean will absorb enough to restore the former 'equilibrium'. Adding to our concern, this uncertainty is doubly amplified by the sure knowledge that our forests are being destroyed, not only removing the trees as sinks or 'buffers' for carbon dioxide, but re-injecting their existing 'fixed' carbon into the atmosphere as new carbon dioxide (Woodwell *et al.*, 1983).

Some of the carbon dioxide removed from the atmosphere by photosynthesis is locked up within the tissues of plants—especially woody plants—under conditions that do not favor the activities of decay organisms. In addition to living trees and shrubs, extensive beds of peat and other organic materials in shallow lakes, bogs and swamps are examples. The carbon content of our fossil fuels (coal, oil, natural gas) originated from just such sources in geologically remote times. Their presence in the earth attests to the fact that production of carbon compounds by photosynthesis processes once exceeded decay, burning and other processes that return carbon dioxide to the atmosphere.

To the ocean's capacity for absorbing carbon dioxide we must add back or replace that which once existed on land, or at least permit nature to do so. This sink could very well equal or rival that of the ocean. It should not be difficult to estimate the capacity of trees to sequester carbon dioxide; one would multiply the estimates of the average volume and density of wood in a typical tree by the number of living, standing trees in the world, a procedure at least as straightforward as many of the estimates obtained in environmental science. Indeed, this has already been done for tropical and subtropical forests (Brown & Lugo, 1984) in a study aimed at estimating how much more carbon dioxide might be released to the atmosphere per year by the ongoing destruction of the remaining tropical forests. An estimate of the worldwide maximum potential quantity of woody biomass might be determined, assuming forests were recognized by humanity as a necessary sink for sequestering carbon dioxide and stopping or reversing greenhouse warming. This potential carbon reservoir is not as great as oceanic and other reservoirs of carbon dioxide, but woody plants provide a greater reserve of bound or 'fixed' carbon than is presently appreciated. Moreover, woody plants exceed crop plants, grasses and herbaceous forbs in the ability to sequester carbon. This is because the crops, grasses and herbs are all processed through digestive, oxidative or decay processes so much more quickly than living or even harvested trees. The contribution of burning firewood to the global burden of carbon dioxide should not be minimized, but living trees are among the world's important sinks or reservoirs for atmospheric carbon dioxide. If this idea could become both folklore and national policy, the danger of runaway greenhouse effects might at least be minimized.

The consummate mitigation may take place unwittingly through conservation of energy or by the use of energy sources other than fossil fuels, such as nuclear energy. But this is unlikely. Rather, the question is, what scenarios of fuel use will minimize the release of carbon dioxide? The answer depends to a large measure on the abundance and usage of the fossil fuels and their costs.

Two scenarios of fossil fuel usage and carbon dioxide emissions were developed by the Institute for Energy Analysis (Rotty & Reister, 1986). The high carbon emission scenario resulted in an atmospheric carbon dioxide concentration of about 1040 ppm by the year 2100. The low carbon emission scenario gave a carbon dioxide concentration of 700 ppm. The high-carbon case would almost surely result in rates of climatic change that would present serious difficulties for humanity. The low-carbon case may allow opportunities for adaptation and shifting of regional agricultural and natural vegetation community patterns.

No one is quite sure whether warmer global temperatures will bring increased precipitation or less. Folk wisdom might suggest that 'warmer is dryer'. This is a good possibility. A warmer atmosphere will undoubtedly be more energetic. It is already established that global climate during the late 1970s to late 1980s exhibited more variability than in any similar period in either recorded or reconstructed climatic history. This 10-year period of warm–cold extremes and high–low precipitation might be expected to recur only once in 1000 years by chance alone, based on past records. The oscillations have been partially attributed to the El Niño warming events of the eastern Pacific Ocean, which altered climate on a worldwide basis during these memorable years. Questions are still being asked about what causes the El Niño event itself. Could it be the beginning of greenhouse gas impacts?

Stratospheric ozone and chlorofluorocarbons

'Good' ozone, a product of photochemical reactions, exists naturally in the upper atmosphere, where it blocks excess ultraviolet radiation from reaching the earth's surface. Solar ultraviolet radiation of wavelengths less than 240 nm is absorbed by both atmospheric oxygen and ozone, but only ozone effectively absorbs wavelengths between 240 and 320 nm. Depending on solar angle and other factors, a 10% reduction in ozone results in a 20% increase in ultraviolet penetration at 305 nm, a 250% increase at 290 nm, and a 500% increase at 287 nm (Cicerone, 1987). Ozone, although constituting less than one part per million in the atmosphere, including the stratosphere, absorbs most of the ultraviolet radiation from the sun. Without this protection, ultraviolet radiation would probably make life on earth impossible. This radiation is energetically capable of disrupting many important biological molecules, including DNA, protein molecules, chlorophyll and other plant pigments. It can increase skin cancer, cause cataracts, contribute to immune deficiencies, and harm crops and aquatic life. Primary production in the surface waters of the ocean can be reduced by excess ultraviolet radiation (Dohler, 1984).

'Bad' ozone, which damages vegetation, oxidizes materials and harms human health, is formed in the lower atmosphere by other, unrelated photochemical reactions involving volatile organic compounds from industry, automobiles and vegetation, and oxides of nitrogen emitted from industry and automobiles. Bad ozone is responsible for more than $1 billion worth of damage to agricultural crops annually and additional extensive damage to trees and natural vegetation.

Since about 1976 the stratospheric, beneficial kind of ozone has been disappearing at certain times of the year. The cause has been linked to compounds known as chlorofluorocarbons, or CFCs, the familiar freons used as refrigerants and aerosol propellants. Freons are good in these roles because they are chemically inert and do not create environmental hazards when used in closed spaces. Inertness, the very property that makes them desirable, is the same property that is ultimately responsible for their potentially harmful effects. CFC-12 (CCl_2F_2) is used in automobile air conditioners. CFC-11 (CCl_3F) is used for blowing the foam used in containers, seat cushions, and automobile dashboard padding. CFC-113 (trichlorotrifluoroethane, $C_2Cl_3F_3$) is used as a cleaning solvent in electronic chip production. On the negative side, CFCs, by decomposing at high altitudes, contribute to the depletion of stratospheric ozone, the protective ozone layer in the region from 15 to 40 km that shields the earth from excessive harmful solar ultraviolet radiation. Among the decomposition products of CFCs are chlorine atoms, which are apparently active in chain reactions that convert stratospheric ozone back to ordinary oxygen, regenerating the chlorine atoms, catalyst-like, and which remain in the atmosphere in the process. Concentrations of CFCs now amount to just over 0.1 ppb by volume of the troposphere, an amount that accounts for almost the whole of the world's production since they were first introduced (Wellburn, 1988).

Besides destroying stratospheric ozone following decomposition, intact

CFCs have a second, equally devastating effect. By absorbing infrared light in the wavelength bands from about 7 to 10 μm, part of the 'window' through which the earth radiates excess heat into space, CFCs act as 'greenhouse gases'. On a molecular basis, CFCs are as much as 10 000 times more efficient than carbon dioxide in absorbing solar energy re-radiated back from the earth that would normally be lost to space (Wellburn, 1988). With the build-up of CFCs, as well as of carbon dioxide, methane, nitrous oxide and other infrared-absorbing gases, this window has become 'dirty'. It is no longer fully transparent to outgoing infrared radiation, and the absorbed wavelengths are gradually heating not only the upper atmosphere but the entire earth as well.

Were it not for these two aspects of stratospheric activity, CFCs would be environmentally harmless. In the lower atmosphere near the ground, they are non-toxic, non-flammable and non-corrosive. These very properties of unreactivity are believed also to contribute to particularly long lifetimes in the atmosphere—as much as 100 years. This gives them plenty of time to diffuse, mix or be actively lifted in storms to higher altitudes, where they can build up in concentration and where their ozone-destroying and infrared-absorbing greenhouse effects come into play.

Why are CFCs reactive at stratospheric altitudes but not near the ground? The normal ozone layer, which absorbs incoming ultraviolet radiation, is part of the answer. Sunlight energy reaching the lower atmosphere has been attenuated by this high-altitude absorption, and low-level CFCs do not receive enough energy to cause decomposition. Once at high altitudes, however, they are bathed in the full intensity of solar radiation, which is capable of disrupting the otherwise inert molecules.

The use of CFCs in aerosols was banned in 1977 in the USA. Possible substitutes for the unreactive freons include CFC-22 (CHF_2Cl) and FC-134a (CH_2FCF_3). CFC-22, already used in stationary refrigeration, is a less stable molecule that decomposes faster in the lower atmosphere and which might have only about a quarter of the ozone-depleting potential of CFC-12. FC-134a is a fluorocarbon that contains no ozone-scavenging chlorine atoms at all. Conversion to these new substances, however, will require several years of engineering effort because they are more reactive with the components of refrigeration equipment, hoses and seals, for example. In addition, FC-134a is still being tested for toxicity and reliable performance.

The first large-scale study of stratospheric photochemistry was aimed at evaluating the possible reduction of upper-atmosphere ozone by exhaust emissions from high-flying aircraft (i.e. SSTs), especially oxides of nitrogen (CIAP, 1974). This was followed by an inquiry into the effects of chlorofluoromethanes on stratospheric ozone (Hudson & Reed, 1979). Beginning in 1976, published accounts of ozone depletion over Antarctica began to appear. The springtime amounts of ozone at altitudes between 12 and 24 km decreased every year to ever lower values, which were expressed in Dobson units. Dobson units are hundredths of a millimeter; they represent the thickness of the layer of ozone that would result if all the ozone in a column directly above an observer at the surface of the earth were brought to standard pressure and temperature (Stolarski, 1988). Measured since 1956, Antarctic springtime ozone levels remained between about 325 and 280 Dobson units until 1979. New record lows in ozone concentration were reported almost annually thereafter. In 1979 the value was 225 units; in 1985, 136 units; and in 1987 only 109 units, which was until then the deepest Antarctic ozone hole yet observed. This reduction began around 1976, forming what has come to be called the 'ozone hole' in the polar atmosphere. An immediate reaction to this discovery was concern by both scientists and the public alike: was ozone also being depleted on a worldwide scale? Apparently yes, according to the Ozone Trends panel in a report (Watson, 1988), especially during the winter and at higher latitudes. Between 1979 and 1987, worldwide stratospheric ozone decreased by about 2.5% between 53 °N to 53 °S, or about 0.35% per year. Polar reductions, which include ozone hole depletion, are as much as 5–30%. Ozone levels above the equator are normally about 260 Dobson units (Stolarski, 1988), which is lower than normal, undepleted levels at the poles. Peak levels reach 450 Dobson units in the late winter or early spring at the North Pole, and 380 Dobson units at the South Pole, the difference being due to atmospheric circulation.

Nitrogen enrichment

Nitrogen is the most abundant component of the atmosphere. Its diverse forms are critical to plant health in the biosphere and energy equilibrium above it. Most atmospheric nitrogen occurs in the gaseous molecular form, N_2, but lesser amounts occur as nitrous oxide (N_2O), nitrogen dioxide (NO_2), nitric oxide (NO), and other oxides of nitrogen, as well as ammonia (NH_3). All of these have a role in the chemistry of the atmosphere as well as its interactions with plant life and the soil. Nitrogen is essential to plant health, but excesses as well as deficiencies can be harmful. The ammonia form in the atmosphere appears to be especially critical.

Oxides of nitrogen

The harmful effects of oxides of nitrogen, especially nitrogen dioxide, have been recognized for many years. Their roles in the formation of ozone in the troposphere, and breakdown of ozone in the stratosphere, are especially critical. Alone, but more particularly in pollutant interactions, they can adversely affect plant growth when concentrations exceed natural background levels.

Considerable work has been conducted to learn the effects nitrogen dioxide, the most toxic oxide of nitrogen, has on vegetation. Studies were reviewed by the EPA in 1981 (US Environmental Protection Agency, 1981) and more recently in the Air Quality Guidelines for Europe (WHO, 1987). Since research has shown that visible injury to vegetation from nitrogen dioxide alone occurs only near a few point sources, acute effects will be treated only briefly here.

The lowest threshold of injury appears to be around 2 ppm for a 4-hour exposure, which is rarely exceeded in nature. Average exposures throughout the growing season would have to exceed 0.25 ppm in order to expect any growth suppression (Ashenden, 1979a). These levels also are unlikely to occur in ambient atmospheres. Even in Los Angeles the highest nitrogen dioxide concentration ever recorded was only 3 ppm, and 200–300 ppb is more typical of polluted days.

The greater direct concern to vegetation lies in the interactions of nitrogen dioxide with other pollutants, in nitrogen eutrophication effects, and in disrupting balances with other nutrients. A more far-reaching consequence may lie with nitrous oxide. While not significant in the troposphere, in the higher-energy stratosphere nitrous oxide contributes to ozone depletion through photochemical reactions, thus contributing to increased ultraviolet radiation. Nitrous oxide also acts as a greenhouse gas, absorbing infrared radiation and contributing to atmospheric warming and hence to global climate. Due to the greater solar energy available in the upper atmosphere, oxides of nitrogen consume ozone there, while near the ground they promote ozone formation in the presence of strong sunlight and hydrocarbons.

Wet and dry deposition of nitrogen oxides are important factors in the redistribution of nitrogen in the environment. Plants have the capacity to remove nitrogen oxides from the atmosphere, most notably by foliage near the top of the canopy, where light intensity and metabolic rates are highest (Bennett, 1975; WHO, 1987). Once in the leaves, they may be metabolized by normal mechanisms and act as supplementary foliar nitrogen fertilizer at low levels. When in excess, they may cause problems of toxicity. Several woody species may be particularly susceptible to injury if their leaves lack the necessary enzymes to metabolize the nitrate and nitrite formed by the absorption of nitrogen oxides (Amundson & McLean, 1982).

There is evidence that low concentrations of nitrogen dioxide below any 'threshold' dose for injury may stimulate growth. This may not always be beneficial, as when plants are made more sensitive to frost, insect attack or diseases. Perhaps more critically, relationships between various plant nutrients may be disrupted, thereby adversely affecting their availability and their equilibrium with nitrogen.

At the ecosystem level, the nitrogen supply is vital. Since nitrogen is a main trigger of plant growth, nitrogen impact to ecosystems means changes in the structure of plant communities. More than 50% of the plant species in central Europe can only compete in stands deficient in nitrogen (Ellenberg, 1987). Plants that are pushed by nitrogen supply to grow fast, dense and high need more cations, such as potassium, magnesium and calcium, for their growth, thus depleting the supply for competing species. Nutrient imbalances follow.

An average hectare of cultivated land in Western Germany is estimated to receive about 200 kg of nitrogen per year from anthropogenic sources. In throughfall of plant stands or at forest edges, nitrogen input can be still higher.

Spruce and pine are particularly adapted naturally to poor, low-nitrogen environments. Stress from excess nitrogen to such species may develop long before the appearance of the traditional responses sought from air pollutants. The high nitrogen content of needles, connected with altered nutrient ratios (e.g. magnesium or boron), may well contribute to altered growth patterns. Inhibition of photosynthetic phosphorylation, as with ammonia, leading to decreases in carbohydrate production, is probably a principal mechanism of toxicity. When nitrite is formed it rapidly breaks down to ammonia and has the same harmful effects as this molecule. Ammonia can saturate the lipids in the cell membrane, increasing permeability.

When ammonia is available, nitrate utilization is impaired. The ratio of reduced to oxidized nitrogen also may have a mechanistic role (Klein & Perkins, 1987). Because coniferous trees have a low capacity to reduce nitrate, its accumulation may provide particular potential stress to these species.

Ammonia

Ammonia (NH_3) is a colorless gas that is readily soluble in water. It has a strong, pungent odor, familiar wherever organic matter is undergoing biological decomposition, such as near stables, manure piles and chicken- and turkey-raising operations, but, most significantly of all, simply normal decay processes.

Ammonia, which is used in many chemical processes, is produced commercially mainly by the direct combination of nitrogen and hydrogen at temperatures of about 500 °C and pressures of several hundred atmospheres in the presence of a catalyst. Although ammonia compounds are sometimes used as fertilizers, ammonia is not easily utilized by plants. Loss of utilizable nitrate ions sometimes occurs in the soil by the action of denitrifying bacteria that convert ammonia and nitrate ions into water and diatomic nitrogen (N_2), which is unavailable to plants. For this reason, growers must take care not to mix fertilizers containing nitrates and ammonium salts, and not to add nitrate fertilizer to compost heaps because they contain ammonia. Atmospheric ammonia probably does not contribute significantly to the total nitrogen supply of a plant, whereas it can be potentially toxic. For example, it is an inhibitor of photosynthetic phosphorylation, which results in reduced carbohydrate production and therefore reduced growth.   Ammonia, although not a major air pollutant, has been cited in local episodes following transport accidents, releases from refrigeration systems, near sources of anhydrous ammonia production, and in areas of intensive livestock management such as near feedlots, mink farms, dairies, poultry and pig farms, and other such sites (Van der Eerden, 1982). Injury to vegetation has been reported to occur usually within 50 m but up to 400 m of such sources.

Finland, which is one of the biggest producers of farmed furs in the world, with some 5600 fur animal farms and a yearly production exceeding 6 million furs, has experienced growth disturbance in forest species (especially Scotch pine, *Pinus sylvestris*) near these fur farms. Probably caused by a combination of high nitrogen levels, nutrient deficiencies and ammonia exposure, symptoms include loss of apical dominance, dieback of the leader shoot, abnormal branch growth and needle fall (Ferm *et al.*, 1987).

The major effects of ammonia occur mostly on the older leaves or needles. Most symptoms resemble those of stress factors similar to those caused by drought, salts, some plant diseases or other pollutants, but symptoms specific to ammonia also exist. These include dark spots or the complete blackening of leaves, rusty marginal spotting of rye and wheat, and dark-brown lesions between the veins on azalea and chestnut leaves, which turn black the next day followed by the entire leaf drying up. Color changes in fruits and vegetable skins and brownish to black lesions and a general browning of fruits also occur. Sometimes plant growth is stimulated; at other times it is reduced by ammonia exposure, depending on the exposure regimes. Stimulation may occur even when necrosis is present. A major effect of ammonia exposure to plants is an increased sensitivity to cold and frost damage. This may be a result of lipid saturation by ammonia in cell membranes, increasing their permeability and decreasing their flexibility (Van der Eerden, 1982).

Regional, more chronic, effects are likely to be of greatest consequence (Soderland, 1981). High levels of ammonium ion deposition in forest ecosystems can lead to nitrogen enrichment—eutrophication—of the soil. Nitrogen deposition in forest stands of the Netherlands may be 10–20 times the natural supply of 5–10 kg/ha per year. One result of this is the encouragement of fast-growing forest species (Heil *et al.*, 1988). The increase of ammonium ion in the soil also appears to inhibit the growth of symbiotic fungi (mycorrhizae), leading to reduced uptake of potassium and magnesium ions by the roots. Moreover, conifer needles that take up ammonium ion from atmospheric deposition compensate for this by excreting potassium and magnesium. These effects often result in potassium and/or magnesium deficiencies, severe nitrogen stress, and yellowing and premature shedding of leaves or needles (Roelofs *et al.*, 1987). These trees become more susceptible to other stress factors too, such as ozone, drought, frost and attack by pathogenic fungi. One study (Roelofs *et al.*, 1985) showed that all of the Austrian pine trees infected with the pathogenic fungus *Sphaeropsis sapinea* had significantly higher nitrogen levels in the needles compared with the non-infected, healthy trees. The nutrient imbalances produced may also have been critical. For these reasons, among others, ammonia (or ammonium) and sulfur dioxide have been described as the two most damaging air pollutants in Western Europe (Heil *et al.*, 1988).

Numerous published fumigation studies suggest that exposures for threshold or no adverse effects from ammonia are 75 $\mu g/m^3$ for a yearly average; 600 $\mu g/m^3$ for 24 hours; and 10 000 $\mu g/m^3$ for one hour (Van der Eerden, 1982).

ETHYLENE

Ethylene (or ethene, C_2H_4) is more than a natural trace chemical in the atmosphere. It is a product of the plants themselves, a growth-regulating chemical critical in senescence, dormancy, fruit ripening and other metabolic processes. No wonder that in elevated concentrations ethylene can be adverse to plant health.

As a natural plant product, trace amounts of ethylene are released from growing plants, but ethylene is also found in natural gas (up to 3%), in the combustion products of coal and wood, in the manufacture of ethylene products (e.g. polyethylene) and in the exhaust of automobiles. Ethylene is a colorless gas with a sweetish odor, that reacts readily with other substances, particularly the halogens. This property is owing to a double bond between the two carbon atoms, which is the hallmark of the so-called unsaturated hydrocarbons. Ethylene is used commercially to ripen green fruits and also as an anesthetic. More significantly, ethylene is released whenever organic substances are burned, as in agricultural burning, incineration and even during the smoking of tobacco. Ambient air concentrations of ethylene have been measured at levels of about 0.03–0.20 ppm, with peaks reaching as high as 1.42 ppm. As a 'greenhouse effect' gas in the atmosphere, ethylene absorbs infrared radiation in a wavelength band extending approximately from 9.8 to 10.42 μm; its absorption strength is somewhat stronger than that of carbon dioxide.

Natural gas-leaks, unburned blow-off gas from oil fields and refineries, and industrial processes that emit ethylene can be sources of sufficient quantities of ethylene to be toxic to nearby vegetation, which can occur at levels of less than 1 ppm to sensitive plants such as greenhouse-grown orchids, carnations, snapdragons, roses, camellias and chrysanthemums (Treshow, 1970).

Ethylene is not directly damaging to plant tissue, but it acts as a growth regulator. In the fruit distribution industry ethylene is used to control the time of ripening of many fruits, such as bananas, which are normally picked green. But for growing plants, when excessive concentrations or exposure during immature growth stages occur, ethylene can upset normal growth and aging patterns. In low concentrations ethylene is important to normal plant growth, where it participates in the regulation of normal maturation processes and in the abscission of leaves, flowers and fruits. Acting as a growth regulator, ethylene plays a role in shoot elongation, fruit ripening and coloration, leaf and flower senescence and abscission, root induction, lateral growth and normal processes affecting dormancy. Excessive ethylene during early growth stages hastens maturation processes and accelerates respiration, causing premature senescence and abscission. Abscission, for example, is normally regulated by a balance between auxin and ethylene in the petiole. Increases in ethylene through pollution upsets this balance, leading to early defoliation (Treshow, 1970).

In the field, near a polyethylene-manufacturing facility where concentrations of ethylene ranged from 0.04 to 3.0 ppm, severe losses of cotton occurred, including plant death within one mile (1.6 km) of the source, and less severe symptoms at more distant sites (Heck & Pires, 1962). These symptoms included leaf abscission and complete defoliation of young plants near the source, seedling death, occurrence of a vine-like growth habit, fruit abscission, deformation of leaves and stems resembling that caused by 2,4-D, leaf puckering, reddening and chlorosis of leaves, loss of apical dominance, forcing of axillary buds and stimulation of flowering, which was, however, offset by the abscission of fruits.

METHANE

Methane (CH_4) is not regarded as a classic air pollutant, but it is another one of those compounds that is important to the natural equilibrium of the atmosphere, and consequently to plant health.

Methane is an end-product of the anaerobic decay of plants. It is the major constituent of natural gas, accounting for up to 97%. It is the dangerous coal-mine gas known as firedamp, and it is also the substance of marsh gas that bubbles to the surface of swamps. In addition to offgasing from oil and gas wells, other sources of atmospheric methane include a substantial contribution from termite nests, the rumen of cattle, goats and sheep, rice paddies and natural wetland ecosystems. All of these sources are mediated by specialized anaerobic bacteria that reduce carbon dioxide to methane.

Methane undergoes only two chemical reactions of any importance: combustion and halogenation. Both of these take place only under very vigorous conditions. An initial flame is necessary to begin the combustion of methane, as in the burning of natural gas. Once started, however, the reaction with oxygen produces enough heat to maintain combustion, producing carbon dioxide and water. Heat or light energy is required for methane to combine with the halogens—fluorine, chlorine or bromine. Methane is unreactive with the remaining halogen—iodine. In the atmosphere, however, where sunlight energy is available, methane strongly reacts with atmospheric hydroxyl radicals (OH) and can be involved in the production of tropospheric carbon monoxide and ozone (Prinn *et al.*, 1987).

Atmospheric methane concentrations, currently between 1.6 and 1.7 ppm, are increasing at a rate of about 1.1% per year, based on measurements taken since 1965 (Mooney *et al.*, 1987). This rate exceeds the rate of increase for atmospheric carbon dioxide. Both are infrared-absorbing 'greenhouse gases' whose build-up is expected to lead to increased average temperatures on the earth in the near future. Methane is about 20 times more effective per molecule than is carbon dioxide in greenhouse forcing. Methane production from the biosphere is also responsive to temperature, and the expected rise in temperature from greenhouse forcing may result in further increases in the rate of methane production. Analyses of ice cores indicate that methane concentrations have already more than doubled over the past 200–300 years, after having remained relatively stable for thousands of years (Kerr, 1983; Stauffer *et al.*, 1985).

SUGGESTED READING

Bolin, B. & Cook, R.B., eds (1983). *The Major Biogeochemical Cycles and Their Interactions*. Published for the Scientific Committee on Problems of the Environment (SCOPE) of the ICSU by Wiley, New York. SCOPE 21. From a workshop at Orsundsbro, Sweden, 1981. xxii + 532 pp., illus.

Cumberland, J.H., Hibbs, J.R., & Hoch, I. eds (1982) *The Economics of Managing Chlorofluorocarbons: Stratospheric Ozone and Climate Issues*, Resources for the Future, Washington, DC (Distributor: Johns Hopkins University Press, Baltimore). xxii + 512 pp.

Khalil, M.A.K., & Rasmussen, R.A. (1987). Atmospheric methane: trends over the last 10,000 years. *Atmos. Environ.*, **21**, 2445.

Mintzer, I.M. (1987). *A Matter of Degrees: The Potential for Controlling the Greenhouse Effect*. World Resources Institute, Washington DC. 60 pp.

Stolarski, R.S. (1988). The Antarctic Ozone Hole. *Scientific American*, **255**, 30–37.

Trabalka, J.R., ed. (1985). *Atmospheric Carbon Dioxide and the Global Carbon Cycle*. National Technical Information Service, 315 pp. (This publication is available as DOE/ER-0239).

Wellburn, A. (1988). *Air Pollution and Acid Rain: The Biological Impact*. Longman Scientific & Technical, Essex. xiii + 274 pp., illus.

Wigley, T.M.L., Ingram, M.J., & Farmer, G., eds (1985). *Climate and History: Studies in Past Climates and Their Impact on Man*. Cambridge University Press, New York (Reprint, 1981 edition). xii + 530 pp., illus.

Woodwell, G.M., ed. (1984). *The Role of Terrestrial Vegetation in the Global Carbon Cycle: Measurements by Remote Sensing*. Published for the Scientific Committee on Problems of the Environment (SCOPE) of ICSU by Wiley, New York, SCOPE 23. From a conference at Woods Hole, Massachusetts, May 1979. xviii + 247 pp., illus.

CHAPTER 15

The Costs of Air Pollution Damage

PUBLIC AWARENESS

In a community economically dependent on a nearby steel mill, how do the residents feel about the pollution it produces? A sociological study asked the question, 'Is there an air pollution problem in your city?' Eighty per cent of those not economically dependent on the steel mill were bothered by air pollution. Only 17% of those dependent on it perceived a problem (Creer, 1968). The point is that whether or not air pollution is a problem is first of all an individual judgement. Personal perceptions and values tend to influence attitudes towards pollution. Costs to the national economy, such as for damage to crops and materials, typically concern us less than what affects each of us personally. The average consumer may not think about crop losses that are reflected in paying a few cents more for a head of lettuce or other produce. We are not concerned with damage to forests until it becomes so severe as to impair their use for recreation or timber harvesting.

What is of concern, and what calls pollution to our attention, is largely that which offends our senses. Impaired visibility, eye irritation, noxious odors or impaired breathing are paramount. These are not only offensive but may impose direct health burdens and economic costs. When one or more of these conditions reaches extreme proportions, action is demanded. Historically, governmental action has been preceded by local or individual action. Examples of such actions date back to the nineteenth century. Historically, farmers whose crops had been damaged by smoke from nearby smelters or other pollutant sources filed law suits to shut down such pollutors. In other cases, farmers were reimbursed for the damages, or sold their land to the pollutor. Such actions were common in Europe and the US throughout the industrial revolution and continue into the 1980s. The experiences gained in defining the extent of the physical damage to the farmer and the associated economic costs of this damage have provided an initial basis for estimating crop losses on a broader, regional scale.

The history of air pollution control suggests that action to control regional or national pollution generally follows after considerable public concern. This happens after the public has become aware of a problem. Awareness may wait until the air has become a thick, off-colored pungent pall. Even then, in a pluralistic society, there are those who may question if the air is polluted or if this is a natural phenomenon.

For example, when smog was noted in southern California after the Second World War, there were those who questioned its origin. Skeptics noted that sixteenth-century explorers had named the bay Bahia des Fumos. Others maintained that the haze simply drifted in from the ocean. But as the smog, eye irritation, crop damage and poor visibility got worse, the scientific community and the public recognized that the cause was not natural and something had to be done.

However, public concern and complaints from the agricultural industry alone were not enough to justify large expenditures for control without first knowing the economic impact of air pollution. Studies were begun to determine the value of losses to crops and ornamentals caused by air pollution. Costs to human health and agriculture were of primary concern, but corrosion effects on materials, soiling effects and reduction in visibility were also considered. All are targets of air pollution.

Economists were interested in both the economic efficiency and equity of controlling pollution—what level of pollution is socially acceptable? What should be done to achieve such concentrations? How can costs be allocated fairly among those who contribute to pollution? The real question is, when does it become beneficial to control pollution, or when does the cost of control exceed the cost of pollution?

Complaints from the agricultural industry, and continuing increases in the number of claimed losses, plus the broader public concern, made it important to determine the losses to crops and other plants caused by air pollution.

The specific air pollutant causing a loss is of concern. Ozone is by far the most significant pollutant in North America and probably Europe. Therefore, emphasis is placed on the economic impacts of this pollutant, although sulfur dioxide is also considered. In the USA, sulfur dioxide is of little consequence. The situation is complicated in Europe, however, since sulfur dioxide concentrations over much of the continent are still high enough to affect productivity. This is especially true in parts of eastern Europe, as with Poland where leaf damage symptoms are still prominent. Where there are co-occurrences of different pollutants, the situation is further complicated.

This is not common enough in the USA to present a problem in estimating economic losses, but it may be in some other countries. However, data are meagre regarding production losses involving multiple pollutants.

COSTS TO AGRICULTURE

Receptors

The costs of air pollution depend on what is affected, i.e. the receptors. Receptors quite logically include much more than the plants with which we are concerned. Visibility and climatological effects, aquatic as well as terrestrial ecosystems, and materials are all receptors. Humans are a major receptor of air pollutants, and much of the cost of air pollution involves establishing safeguards against impaired human health.

Air pollutants are intimately associated with a number of respiratory diseases, largely in aggravating existing disorders such as emphysema, bronchitis and asthma. The costs of these effects to individuals and society are considerable. Our more immediate concern here, though, is with plants.

Data bases and needs

How are estimates of economic costs obtained? What data go into calculating economic cost estimates? First we have to know the effects that air pollutants have on plants, and the pollutant concentrations that cause a given effect. Then we must know the land area of the crops affected, the normal production levels obtained, and the demand for the affected crops. Information on the demand for a crop is necessary in order to determine how the increased production resulting from less pollution might influence the price. Changes in price in turn may affect the decisions farmers make concerning how much of each crop to plant. Economists integrate such data into a model of the crop supply and demand. These models are designed to estimate the economic losses of air pollution, or, putting it somewhat differently, the economic benefits of controlling pollution to varying degrees. If done correctly, each cost or benefit estimate should be in units that represent changes in income for farmers or buyers.

The accuracy of the estimates used to predict the costs of air pollution can be no better than the available data. The closer experimental procedures are to real world conditions, the more useful the data are to making accurate economic assessments. The lack of necessary data on plant response to air pollution for use in economic assessments has been a serious obstacle to many attempts to quantify the costs of air pollution. Specific biological response data, linking pollutant levels and crop yields, provide the basis for most cost estimates. Several approaches have been applied to develop cost estimates.

Regional strategies and impacts

Many of the economic assessments in the literature focus on losses in specific geographical regions and concern specific crops (Moscowitz *et al.*, 1982; Benson *et al.*, 1982; Manual *et al.*, 1981; Shriner *et al.*, 1982; Leung *et al.*, 1982; Page *et al.*, 1982). Generally these studies looked at losses to either producers or consumers, but not both. Some failed to adequately model the farmers' or buyers' behavior. Also, many of the interactions among crops of varying sensitivity were not incorporated into these assessments. Estimation procedures that ignore potential shifts in crops caused by pollutant damage could result in significant errors in estimating agricultural costs from increases in air pollution levels, or benefits from reductions (Smith & Brown, 1981). Most significant from a biological basis, dose-response data available in earlier regional estimates were simply inadequate to fully understand the amount of crop loss from a given exposure. While results of regional assessments cannot be applied on a national scale, they do reflect extreme cases in regions where pollution is especially intense or a pollutant-sensitive crop is dominant.

Some of the early figures do give a rough idea of the monetary losses to agriculture that were attributed to regional air pollution. Smog was reported to have cost Los Angeles County $500 000 already in 1949 (Middleton *et al.*, 1957). By 1953 the estimate was $3 million; and in 1956 losses were reported to have reached $5 million in Los Angeles County and over $1 million in the San Francisco area.

Early estimates of economic losses due to oxidants and sulfur dioxide were largely subjective and lacked any credible yield loss data. Estimates were based mostly on visible symptoms. For instance, the National Academy of Science (1977) placed losses in California at $8–10 million in the 1960s and on the east coast at $15 million. These figures later were rather arbitrarily raised to $500 million in the US, based on the acceptance of yield effects without visible symptoms and recognition of additional sensitive species (US Department of Health, Education and Welfare, 1966).

Actual dose–response data often were lacking. In one study, for instance (Leung *et al.*, 1982), crop yield reductions were based on the difference between actual yields (at 1975 ozone levels) and yields predicted by the linear response functions (at zero ozone concentration). Yield adjustments were predicted from these functions and used to estimate changes caused by ozone. Two indicators of impact were calculated: producer surpluses and consumer surpluses. These are economic indicators that should be explained. A 'producer surplus' is the difference between what the grower actually receives for his crop and the minimum that he is willing to accept. Thus, a pollutant may reduce yields to a degree where the money received from a crop is not enough to warrant growing it. A reduction in 'producer surplus' represents a loss to the grower.

A 'consumer surplus' is the difference between what the consumer is willing to pay and what he actually has to pay. Thus, if a pollutant causes a loss in yields, the price of the crop would be expected to increase and the 'surplus' reduced. A reduction in 'consumer surplus' represents a loss to the consumer since the price he has to pay rises and comes closer to what he is willing to pay, or even might exceed it. The yield study showed that the yield losses caused by ozone reduced both the producer and consumer surpluses.

Some studies have attempted to circumvent the lack of dose–response data by using actual field data on yields across a geographical area. For example, one regional study based loss estimates on county production records and compared these with ozone concentrations in those counties (Mjelde *et al.*, 1984). This method was employed in analyzing the effects of ozone on Illinois grain farms. The direct economic consequences on farmers' income were measured, as well as the physical crop loss. In other words, the relation between yield, profits and ozone across many farms was studied. Generally, ozone had a negative and significant impact on profit. It amounted to a loss to Illinois farmers of about $50 million in 1980.

The California Air Resources Board sponsored an imaginative and ambitious program to estimate the economic impact of ozone in the San Joaquin Valley—one of the richest agricultural regions in the country, with over $5 billion (i.e. $5 thousand million) in agricultural annual receipts (Rowe & Miller, 1985). The economic model used estimated losses incurred by producers and consumers. Many of the most important crops grown in the San Joaquin Valley, notably cotton and grapes, are highly sensitive to ozone. Other major fruit and nut crops and vegetables have varying degrees of sensitivity. Thus, the potential for large economic losses from air pollution exists. Ten crops were selected for detailed analysis and 31 for the final economic analysis. The analyses accounted for changes in farming practices and market conditions caused by changes in yield from air pollution. Both sulfur dioxide and ozone were considered.

A regression approach was used to study pollutant effects on crop yields. The baseline ozone and sulfur dioxide conditions were first determined at each location in the San Joaquin Valley where data were available. The hours that ozone concentrations exceeded the California state standard of 100 ppb, over 80 ppb, or over 120 ppb were determined. It is useful to note, with regard to understanding dose–response effects, that during the growing season of 1981, for example, there were 717 hours when ozone readings in the valley exceeded 100 ppb and 203 hours that exceeded the national standard of 120 ppb. In 1978, the federal standard was exceeded over 300 hours. In some years, 200 ppb has been exceeded. The relationship between these ozone and sulfur dioxide concentrations and county yield records was then analyzed by regressing yields on pollution levels and other explanatory variables. Chamber study data for the crops being studied were reviewed to compare against the regression data.

A mathematical programming model of California agriculture was applied to correct the estimates for how farmers would change the amount and mix of crop acreage planted due to changes in yields. The resultant effect on production costs also is predicted by this model, as well as how changes in production affect prices and quantities sold.

The yield regressions for these data found statistically significant effects of ozone on yields of dry beans, cotton, potatoes and grapes. These results were consistent with those expected from the review of chamber studies. Results suggested that ozone was also causing yield losses in 16 of the crops. Sulfur dioxide effects were found only for potatoes grown during the winter in one area.

Improvement in yields would occur with improved air quality. Although higher yields might result in lower prices to the growers, the total benefit from the increased tonnage is positive because of large gains to consumers and some producers. Also, lower prices may be offset by reduced costs associated with the increased yields—the same production can be obtained from less cropped land area, which means less plowing, less fertilizing, less pesticide use, less fuel consumption and less time required to manage.

An interesting relevant interaction occurs with grapes. California produces the bulk of US grapes and hence has a major influence on price. So when yields increase owing to reduced ozone, the same tonnage can be produced with fewer acres at less cost. Increased yields of grapes would lead to reduced prices that would have greater benefits to the consumer.

California production has a less significant impact on cotton prices, so much smaller acreage reductions occur relative to yield increases. Both producers and consumers would benefit from pollutant control, but not necessarily proportionately. Yields of cotton, for instance, would increase considerably under reduced ozone, but this has little effect on national prices or prices to consumers.

The benefits of air pollution reductions to producers in the San Joaquin Valley were estimated to be $31.8 million if ozone concentrations were kept below 120 ppb; $82.9 million if below 100 ppb; and $92.5 million if below 80 ppb. Benefits to producers and consumers combined would be $42.6, $105.9 and $117.4 million, respectively, for the three ozone scenarios. The approach used here worked well in identifying losses in the more sensitive crops in the area but was limited in its ability to detect possible changes in less sensitive crops.

The San Joaquin study emphasized the importance of using an economic model that accounted for both market and farm reactions. This approach provides policy and decision makers with realistic data on which to base air quality standards since the costs of achieving a certain air quality could be measured against the benefits that may be achieved. For instance, excellent monetary benefits are realized by reaching a 100 ppb value, whereas far less is achieved from the more stringent 80 ppb.

National strategies

Despite many limitations, preliminary efforts began in the 1960s to estimate losses on a national basis (Benedict *et al.*, 1973). In a pioneer study conducted by the Stanford Research Institute, both laboratory and field data from controlled exposures were utilized. Since oxidant concentrations were not well monitored at the time, however, dose–response data were lacking. Consequently, the loss models were based more on the presence and concentrations of ozone precursors, including hydrocarbons, than on production data. Potentially plant-damaging concentrations of ozone, PAN, oxides of nitrogen, sulfur dioxide and hydrogen fluoride were determined by county throughout the USA. Damage was based on estimated ozone concentrations that exceeded the 120 ppb air quality standard.

The dollar value of crops and ornamentals in these counties was determined. The value of each crop was multiplied by the percentage loss expected under each pollutant condition, based on current crop prices and dose–response data. National-level losses were summed for state, region and county. The total annual dollar loss to crops in the USA was calculated to be about $87.5 million, of which $77 million was due to oxidants, $4.97 million to sulfur dioxide and $5.25 million to fluorides. The total loss to ornamentals was about $47 million. Economic losses to vegetation in 1969 were estimated to be about $125 million. The study was subsequently updated utilizing a wide range of dose–response data, increases in crop values and better air quality data (Benedict *et al.*, 1973). The more complete crop coverage caused loss estimates from ozone damages to be raised to between $1.3 and $3 billion. No account was made of benefits to consumers (Benedict *et al.*, 1973; Ryan, 1981).

The early SRI study provided a background on which to build. However, national estimates utilizing the most recent data and more valid economic models were necessary in order to justify the setting of National Ambient Air Quality Standards. To provide such information, the US EPA set out to determine the potential economic losses attributed largely to ozone over the entire USA for the major crops. The National Crop Loss Assessment Network (NCLAN) was established. Through this program, scientists have made considerable progress in combining ozone exposure data (i.e. dose–response data) with production of a number of important crops. Major agricultural crops were exposed to ozone (as well as sulfur dioxide and nitrogen dioxide in some cases) at ambient and higher concentrations in open-top chambers in the field. Although it could be argued that even these conditions are not the same as actual field conditions, the method is generally accepted as realistic. Results of these studies provide the basis for economic assessments largely of ozone impacts. Such models should enable assessment of the national consequences of ozone pollution on agriculture and provide a basis for policy-making legislators.

Emphasis has been placed on the six major crops that account for over 75% of the US cropped land area (US Department of Agriculture, 1982). The crops are corn, soya beans, wheat, cotton, grain sorghum and barley. The effects of ozone on yields was studied at a number of sites over the country, representing different environmental conditions, over a period of years. These data were then used to assess the economic benefits and costs arising from changes in ambient ozone concentrations (Adams *et al.*, 1986). The economic models were adjusted for uncertainties, such as other environmental stresses (e.g. water stress), where possible. An even more critically uncertain economic parameter was the agricultural export sector. An export sector can have a strong influence on the demand for US products, and this was included in the model.

The acreage of one crop in relation to others was also considered, because an increase in one crop's acreage would be expected to be accompanied by a change in the yield of other crops. The model provided a basis for assessing the economic effects of changes in rural ozone concentrations. But before this, model prices, quantity and acreages were compared with results from 1980 production data and values.

NCLAN dose–response data were incorporated in the economic model. Four ozone concentration scenarios were considered: 10%, 25% and 40% reductions in ozone, and a 25% increase. The changes were measured as departures from the 1980 actual ambient ozone levels. The 10% and 25% values were chosen because of their potential importance in standard-setting policies. They parallel alternative ambient one-hour maximum air quality standards of 100 and 80 ppb that have been proposed. The 40% ozone reduction reflects background or even lower concentrations.

NCLAN data show that with a 25% ozone reduction the respective average yield increases would be: cotton, 9%; soya bean, 6.5%; wheat (winter) 3.4%; wheat (spring), 1.5%; corn, 1.2%; grain sorghum, 1%; and barley, 0.2%. It may not sound like much until one considers the huge land area and millions of dollars represented.

Translated into dollars, savings of $0.7 billion dollars would be realized from a 10% reduction in ozone; $1.7 billion from a 25% reduction and $2.5 billion savings from a 40% reduction in ozone concentrations. If ozone were to increase 25%, the added cost to agriculture would be $2.1 billion (Adams *et al.*, 1985).

In the final assessment, the NCLAN study utilized seasonal ozone averages based on 8-hour daily average ozone concentrations averaged over a 3-

month season (Adams *et al.*, 1988). It also adjusted for moisture stress conditions and updated the ozone and price data. The economic benefits were generally similar to previous estimates (Table 15.1).

As shown in Table 15.1, consumers under two scenarios benefit more than producers, but in all cases both benefit. The benefit to domestic consumers comes from lower prices for several of the commodities. The increase in producers' surplus (the difference between what the producer actually receives and the minimum he is willing to accept) arises because of the complex interactions of the demand and supply relationships within the model. Almost all geographical regions benefit from reduced ozone and suffer losses from increased ozone. Regions where most of the crops are produced (e.g. the corn belt) or where ozone levels are highest (e.g. the Pacific coast) benefit the most from ozone reductions. Foreign consumers realize a large relative percentage change in surplus. About 20–46% of the annual consumer benefits accrue to foreign consumers.

Agricultural yield benefits of ozone reductions in other studies show much the same. In one study, evaluation of changes in crop yield from eliminating all pollution, multiplied by a constant market price (from a damage function analysis), showed benefits from $2 to $5 billion (Shriner *et al.*, 1982). Since the damage function was simply a composite of the individual commodities without considering interactions, this was not a true economic estimate. Also, the response data represented only a single year of NCLAN data. In another study the benefits of ozone reduction, as measured in economic surplus, were estimated to be $1.2 billion (in 1978 dollars) (Kopp *et al.*, 1984) (see Table 15.2).

Table 15.1. Economic benefits from alternative ozone scenarios[1]

Seasonal standard[2] (average ozone conc. in ppb[3])	Producer's benefit	Economic Benefits (US dollars, in millions)			Total economic effect
		Consumer's benefit			
		Domestic	Export	Total	
60	414	158	40	198	612
50	769	531	374	905	1674
40	818	985	847	1827	2645

[1] Adapted from Adams *et al.* (1988).
[2] Assuming 95% probability of not exceeding the standard of a 3-month average of daily 8-hour averages.
[3] ppb (parts per billion) in American usage equals parts per thousand million.

Table 15.2. Some recent estimates of national economic consequences of pollution

Study	Crops	Annual benefits of control (billions of dollars)	Ozone reduced to
Stanford Research Institute (1981)	Corn, soya beans, alfalfa and 13 other annual crops	$1.8	120 ppb[1]
Shriner *et al.* (1982)	Corn, soya beans, wheat, peanuts	$3.0	25 ppb
Adams & Crocker (1984)	Corn, soya beans, cotton	$2.2	
Adams *et al.* (1984a)	Corn, soya beans, wheat, cotton	$2.4	40 ppb
Kopp *et al.* (1983)	Corn, soya beans, wheat, cotton, peanuts	$1.2	40 ppb
NCLAN (Adams *et al.*, 1988)	Cotton, wheat, soya beans, corn, sorghum, barley	$0.7 $1.7 $2.5	10% 25% 40%

[1] ppb in US usage is the same as parts per thousand million (e.g. 120 ppb = 12 pphm = 0.12 ppm).

Despite the different approaches, the conclusions of these studies are sufficiently similar to suggest that annual agricultural losses in the USA are on the approximate order of $2 billion. However, except for the NCLAN study, most of these estimates do not consider the mitigating effect of moisture stress, especially in non-irrigated crops. The interaction of low moisture may reduce the benefit estimates by about 25% for each ozone assumption (King, 1987).

The estimates from all of these studies are sensitive to a number of uncertainties; these include both biological and economic factors. For example, a lower monetary exchange rate would also reduce the benefits in export markets due to the commodity being worth less. This would partly offset the gains from increased production. Switching of crop cultivars, the inclusion of US farm program provisions, and changes in commodity demand will also affect these estimates. Also, it should be noted that the benefits are based on reductions in ozone levels, not on actual ozone concentrations. Reductions may be potentially great near urban centers but less likely in more remote areas where the crops are grown. Ozone reductions are not necessarily equivalent to actual concentrations. Finally, it may not be realistic to presume that ozone concentrations can be reduced to 40 or 50 ppb for a growing season average at any cost, since these values are already close to background.

BENEFITS OF CONTROL

Ozone clearly has an economic impact on the agricultural sector of the economy. But agriculture is only one sector. When health, materials effects, esthetic values and other vegetation also are considered, agricultural benefits contribute approximately half of the total economic benefits of ozone control (US Environmental Protection Agency, 1986).

The NCLAN program has provided a good basis for estimating economic losses to the more important agricultural crops. However, there are important sectors involving plants that have not been addressed. Little is known about the economic effects of ozone or other air pollutants on other than agricultural ecosystems. Nurseries and orchards could fall into this category, together with rangelands and pastures, urban parks, ornamental plantings, home gardens, forests and national parks. With but a few exceptions, little is known about the possible effects of any on these systems.

In local situations, as where smelter emissions have killed large areas of forests, or in southern California where ozone is well documented to have killed large areas of pine trees, some economic data are available. But this does not provide much insight as to the possible long-term effects on timber production when the damage is less pronounced. Good economic models must be capable of capturing the loss in timber volume produced and integrate the longer period of time required for trees to reach a merchantable size.

Where natural ecosystems provide a free public service, such 'free' goods might theoretically be viewed as valueless. Realistically, quite the reverse is true, hence there is no agreement as to the value of such systems to society, and there are limited economic data (Farnworth *et al.*, 1981). Such values could be based on replacement costs, such as the cost of revegetation following a forest fire, or on commercial value, such as the value of timber in the market place. Natural ecosystems, including forests managed as wilderness areas and national parks, provide both priced and unpriced benefits to society. But they are difficult to quantify.

Losses in recreation and other benefits from forests are extremely difficult to evaluate. Much has been written about various forest declines, but productivity and economic losses have scarcely been addressed (Treshow, 1988).

Benefits of control must be weighed against the costs of control. The costs of ozone control involve largely controls on automobile and truck emissions, as well as on such stationary sources as power plants, to reduce the nitrogen oxide precursors. For instance, in 1989 the US EPA proposed requiring automobile gasoline to be refined to achieve a lower vapor pressure of the product. This strategy aims at reducing the amount of hydrocarbon vapor escaping into the atmosphere, thus reducing the amount of precursors that participate in photochemical ozone production. The offsetting considerations include increased costs of producing such low-volatility gasoline, higher prices at the pumps, and increased difficulty starting engines in cold weather. Human 'costs' of air pollution involve such intangibles as the inconvenience of non-starting engines or the clumsy handling of the two-hose gasoline delivery system used in California to capture gasoline vapors displaced from the gas tank during filling.

The costs of reducing ozone from the present 120 ppb hourly maximum to a proposed 80 ppb are estimated between $2 and $3 billion. When all the receptor sectors (e.g. health) are considered, the benefits clearly appear to exceed the costs of control, at least in the USA.

SUGGESTED READING

Adams, D.D. & Page, W.P., eds (1986). *Acid Deposition: Environmental, Economic, and Policy Issues*. Plenum, New York. xii + 567 pp., illus. Based on a conference at Plattsburg, New York, 1983.

Freeman, A.M., III (1982). *Air and Water Pollution Control: A Benefit–Cost Assessment*. Wiley–Interscience, New York. xiv + 186 pp.

Izaak Walton League of America, Arlington, Virginia (1982). *Effects of Air Pollution on Farm Commodities*. Proceedings of a symposium at Washington DC, February 1982. xii + 176 pp. illus.

CHAPTER 16

Air Quality Standards

In 1961 a conference was held at the University of Michigan. A then-curious question was posed. Should the government set standards for the quality of air that would be required to protect human health and the environment? Two views were dominant: yes, we needed to set standards to prevent further deterioration of the air; and no, standards were not the best approach and would be unrealistic. Now, air quality standards are an accepted, integral part of the government environmental protection program. Yet, the second contingent had a point. The standards are yet to be met.

Even this degree of action was precipitated only after decades of concern. Prior to the 1960s, pollution mostly was associated with industries, alone or collectively. Pollution in the cities typically came from these factories, and the people affected often worked in them. Complaints were few and the smoke went unabated. There were other reasons, of course, for the incessant delays to controlling pollution. It was as much a matter of circumstances as lethargy.

We shall take only a cursory look at the long history of urban smoke. Since the beginning of the industrial era in North America and Europe, pollution has become increasingly more intense and threatening to both human and plant health. Early in the twentieth century, a number of the most polluted cities passed ordinances encouraging 'smoke abatement'. Smoke, sulfur dioxide, soot and ashfall were measured, but the technical steps needed to reduce the smoke followed very slowly. Controls were costly to the polluters, and the pressure from government was rarely strong enough to mandate effective action.

Progress was made in the USA over a period of decades, but several events thwarted the best of intentions. After an encouraging start in studies to monitor pollution, and legislation to control it in many industrial cities in the early 1900s, along came the First World War. Any priority to curb pollution was lost. Economic conditions after this, including the depression

210

years of the 1930s, also reduced concern for controlling pollution from the few remaining industries. Then came the Second World War. Already it was 1946 before the pollution situation again began to attract serious concern.

The situation was much the same in Europe. In Great Britain, coal that had made England pre-eminent in the Industrial Revolution also led to its greatest pollution problems. Pollution was aggravated further by the absence of oil or gas, which helped mitigate pollution in the USA. The wars were obviously far more disruptive in Europe, and it took major pollution episodes in the 1950s to stimulate action for a Clean Air Act in Britain in 1956.

A major thrust towards pollution control in the USA arose with the recognition of photochemical pollution, beginning in the 1940s. Although smoke pollution and damage to crops and forests had occurred near urban centers around the world for decades, it was just as much the pungent, irritating haze, eye irritation and reduced visibility in the sunny skies of southern California that aroused the public concern.

Decades of air quality measurements had identified the pollutants that caused the most serious effects on human health and agriculture as well as recognizing many of the pollutant sources. The damage caused to crop and human health also was recognized. Now it remained to control these pollutants.

Prior to the application of governmental regulation of air pollution, the public essentially had but one recourse—filing a civil suit against the offending source. Action for damages was not always unrewarded. But neither was remuneration or pollution control readily achieved by litigation. Such civil suits fall under tort laws, which deal with harm to persons and/ or property. While such legal action attempts to redress that harm, it generally does not abate the problems. But lawsuits at least tended to reduce them.

Tort law denoted that pollution could constitute a nuisance owing to its interference with the use and enjoyment of one's property. Much the same purpose is served by trespass suits, but these are applied more when there is a physical invasion of a property owner's land. For damages to be awarded though, some damage generally must be established; and there is little mandate to remove the pollutant. In other words, a successful suit against an offending pollutor could renumerate the plaintiff, or suing party, for damages, but it did not stop the pollution. Abatement followed largely when an injunction was served against the polluter to stop releasing pollutants. An indirect control also came when the cost of litigation exceeded that of control.

Despite the monies often awarded under nuisance or trespass provisions of tort law, pollution too often continued unabated. Tort laws were not an adequate remedy for pollution. Specific regulatory action was necessary.

Before the Second World War, control ordinances were considered to be city problems and were covered by city ordinances to protect health and welfare. Significant local legislative action was motivated by the rapidly increasing photochemical smog problem in the Los Angeles Basin. The Los Angeles County Air Pollution District was formed and given a mandate far broader than the old urban smoke ordinances of eastern cities. Most significantly, the legislation set a precedent of concern that extended beyond simply smoke. In 1947, the responsibility was extended to the States when California enacted the first such air pollution legislation. The federal government first entered the air pollution field in 1955 with the passage of the Air Pollution Control Research and Technical Assistance Act. The Act provided funding but no regulatory authority.

Authority to grant funds to state and local control agencies was provided in the 1963 Clean Air Act, and in 1967 federal legislation required every state to have a state-level program. However, it failed to produce appreciable results.

Several strategies have been utilized: they include emission standards, air quality taxes, or economic incentives, total emission strategies and ambient air quality standards (Godish, 1985). The approaches used in the USA and among European countries often differed in basic concepts despite the similar goals (Edelman, 1968).

POLLUTION CONTROL STRATEGIES

Emission standards

In this strategy, limits are set on the amounts of emissions allowed from a given pollution source. The best practical means, or best available control technology, is applied to limit the emissions. This has been one of the earliest approaches to air pollution control, and is also the most direct (Roberts, 1984). It is common in such countries as the UK, the FRG and the Netherlands (Hartogensis, 1979). The UK passed laws to prohibit the operation of the most offensive factories early in the nineteenth century. This was not always practical, as when the products of the factory were needed, and the law was soon modified to place the factory under a specially created government-controlled inspectorate. These 'alkali inspectors' acted as adversaries promulgating any control methods that were developed; thus local authorities were empowered to control emissions from furnaces and any new industries. This early history may have led in the UK and Commonwealth countries to the characteristic cooperative approach between polluter and the government agency.

In the USA, air quality standards provide the basic approach for certain pollutants, but the emissions approach is used for hazardous pollutants such as beryllium, asbestos or mercury, and in the 'New Source Performance Standards'. Known as smoke ordinances, local governments became involved in smoke regulation in the 1800s. While the black smoke so pervasive in Victorian-era cities surely constituted a nuisance, the effects on health at first were debatable. But over a period of decades health effects became of increasing concern, and by 1912 most of the large cities had smoke control laws. It took a Supreme Court ruling in 1916, however, to establish that such laws were constitutional and legally valid in the USA (Stern *et al.*, 1986).

Whether referred to as smoke ordinances or an emission standard, the purpose remains to regulate the maximum amount of a pollutant allowed to be released into the atmosphere from any source. Measures used range from pounds of particulate per million BTU heat input, to grams of carbon monoxide per vehicle-mile, or other values depending on the source and pollutant.

Standards also have been placed on the type of control equipment that must be used. This can refer to a heat source equipped with precipitators of a specified degree of efficiency. Wording may be vague, such as 'best available control technology' or 'economically feasible'. This leaves the local control agency to determine if a source is providing the best available technology for the money spent. The procedure is usually to ask the polluting industries to submit several different control strategies together with the annual cost per ton of pollutant removed for each. The regulators are then able to make a judgement about costs relative to benefits of the control strategies. Often painful negotiations follow in which the industry will claim certain economic demise under some strategies while the regulators cite inflexible emission standards or other reasons for requiring a certain (often expensive) control strategy. In the USA, state air quality agencies are constantly searching for guidelines for choosing effective control strategies without imposing undue economic hardships on their industries. The balance is often precarious. It would not be politically or economically wise to shut down a job-providing, revenue-producing industry just for the sake of reducing some particulate or gaseous emissions by a fraction—a tiny fraction—of the state-wide or county-wide emission inventory. Gradually, rules of thumb have led to regulatory attitudes in which an annual cost of about $8000 to $14,000 per ton of pollutant removed is considered justifiable (equipment and operation costs combined). Many control strategies exceed this value, of course, and many can cost less.

The economic strategy

In a few instances, a tax is placed on the emitters that is based on the amount of pollutants emitted. The idea is that taxes are set high enough that it would pay the polluter to control the emissions. Variations of this include incentives to reduce emissions such as tax rebates, public loans or subsidies for control equipment. Such incentives have also been combined with other strategies.

Total emission strategy

This strategy has evolved since it became evident that air pollution was a regional, not just a local, problem. The idea is that the total emissions for a region be limited, even it it means no more industrial growth. With growth limited, new industry must be non-polluting or there must be trade-offs by further controlling existing pollution. New potential polluters are encouraged to locate in less polluted areas. This may not effectively control total pollution, but it does make it more feasible for smaller regions to achieve air quality standards. The principle of non-significant degradation or deterioration, accepted in the USA in 1973, utilizes this type of strategy.

Air quality standards

Emission and equipment standards play an important part in abatement processes, but in the 1960s a fresh approach emerged. This was the application of air quality standards. Standards get directly at the quality of the ambient atmosphere. The philosophy was that, at a national level, it would be possible to develop information, i.e. criteria, that would indicate the concentrations of a pollutant that would be harmful. A 'standard' could then be based on these criteria. Air would not be permitted to have more than an acceptable amount of such pollutants. In many regions the 'standard' is more realistically a 'goal', still far from being achieved.

The Clean Air Act of 1970 gave the US Environmental Protection Agency the authority to establish a nationwide program of air pollution abatement and air quality enhancement. Air quality was defined for certain pollutants, the 'criteria pollutants'; and, based on these criteria, National Ambient Air Quality Standards (NAAQS) were to be established.

APPLICATION OF STRATEGIES

To almost everyone's surprise, the National US Center for Air Pollution Control (The parent, pre-EPA agency) published the first criteria document in 1967 shortly after the enabling legislation. It brought together virtually all the scientific information on the effects of sulfur dioxide. This efficiency proved politically unpropitious. The very low level of sulfur oxides considered acceptable—an annual mean of 15 ppb (40 μg/m^3)—raised considerable concern. The 1967 federal act had spelled out the criteria–air quality standards approach. However, the first criteria document was recalled to be re-evaluated after appropriate consultation with industry. The document was then reissued with new values in the form prescribed by the new law.

By 1976 criteria documents had been developed for the six 'criteria' pollutants (photochemical oxidants, carbon monoxide, particulate matter, sulfur oxides, nitrogen oxides and non-methane hydrocarbons) designated in the Clean Air Act of 1970. Both primary and secondary standards were promulgated the next year. A seventh criteria pollutant, lead, was added in 1977.

Primary standards were meant to protect public health and the less stringent secondary standards to protect public welfare from 'any known or anticipated adverse effects'. This was defined as including, but not limited to, effects on soils, water, crops, vegetation, man-made materials, animals, wildlife, weather, visibility, climate, property and hazards to transportation, as well as effects on economic values and personal comfort and well-being.

The 1970 amendments basically set uniform National Ambient Air Quality Standards, designated Air Quality Control Regions, and required State Implementation Plans (SIPs) to achieve the standards. They set emission standards for new or modified sources and for hazardous pollutants, and established federal enforcement authority in case of air pollution emergencies or inter-state and intra-state air pollution violations.

The next few years showed some progress in regulation, but modifications still were needed for successful implementation. Congress again amended the Clean Air Act in 1977. The amendment postponed compliance deadlines for national primary air quality standards; postponed, and in some instances modified, federal automobile emission standards, including the concept of prevention of significant deterioration (PSD); and provided a mechanism for flexibility in regulations that would still allow industrial growth.

Amendments in 1977 required that all existing criteria documents be reviewed at 5-year intervals by a newly created Clean Air Science Advisory Committee (CASAC) (Padgett & Richmond, 1983). The Act also directed the EPA Administration to complete reviews of all existing standards and criteria before the end of 1980 and at 5-year intervals thereafter. Standards were to be revised as appropriate, based on the reviews.

Once the criteria documents are prepared by the EPA and reviewed by the CASAC for scientific content, public hearings are held to provide further input. Revisions may be needed and a second public hearing held, but following acceptance, an EPA Staff Paper, or summary identifying critical elements, is prepared and a 'closing' report is sent to the administrator of the EPA. Based on the scientific conclusions, the administrator then is required to set the final air quality standards. The Amendments to the Clean Air Act did not place any limits on the time allowed to set a standard. This can be a major weakness in the legislation when an EPA administrator is reluctant to make a commitment.

Another problem is that the EPA administrator is not obligated to follow CASAC recommendations. For instance, in 1988 after reviewing the new sulfur dioxide criteria document, a new 1-hour limit for sulfur dioxide emissions was recommended based on sensitivity of asthmatics. EPA considered that the health effects were relatively minor and current regulations were adequate to protect public health. If costs of control are not considered, the question becomes philosophical. Should standards be set to protect the health of the entire population, or only healthy individuals?

Delays occurred in setting ozone standards when it developed that plants were more sensitive than humans, thus requiring that the secondary standards be set more stringent than the primary. This contingency has yet to be resolved as of 1988, although recommendations to change from a 1-hour standard to a 3-month average of daily 8-hour averages may resolve the situation. This type of standard is far more realistic for plants.

Congress recognized that there was no sharp threshold delimiting when adverse effects first became evident. It intended that the administrator exercise judgement in setting the precise standard. A major difficulty in doing this is to find a single value that can be agreed on as 'adverse'. The term 'adverse' itself is difficult to define, and even when defined so many underlying environmental conditions influence effects that one value is difficult to delimit unless it applies under the theoretically most sensitive combinations of conditions.

Achieving the air quality standard

Setting national ambient air quality standards (NAAQS), complicated as the process has become, is still much easier than achieving these standards. Once an ambient standard is promulgated, each state then is required to prepare and submit a State Implementation Plan (SIP) to the EPA for approval. The SIP must denote how the state will proceed to achieve the primary standard. The SIP approach sounds workable but there are problems. Any new pollutant source may cause the NAAQS to be exceeded. To minimize such inputs, New Source Performance Standards (NSPS) must be met. Existing facilities also may be required to be upgraded to meet such standards. The NSPS regulation must be met by using the best available control technology (BACT).

The major problem arises with automobiles. These and aircraft operate on an inter-state basis and therefore are directly subject to EPA regulations unless a state demands more stringent controls on their emissions than the federal regulations require. The main problem, however, lies in applying controls to so many individual pollutant sources, and the fact that the most serious automobile-related pollutants, such as ozone, are formed secondarily in the atmosphere by the action of sunlight.

Controls and regulations of stationary sources are relatively clear-cut compared to those on mobile sources, notably the automobile. Pursuant to the Clean Air Act of 1970, the original maximum allowable 1-hour average of 80 ppb ozone, not to be exceeded more than once a year, placed virtually every major US city in a non-attainment status. Objections and litigation by opponents of such a standard argued successfully for a less stringent standard, and the EPA relaxed the standard to 120 ppb. Data now available indicate that even the 80 ppb may have been too high to protect all vegetation. In any case, increasingly more stringent emission controls were needed to achieve even the higher value.

National emission standards for light-duty vehicles have been increasingly tightened since 1967. The allowable emissions for 1981 vehicles and subsequent model years are 0.41 g hydrocarbons, 3.4 g carbon monoxide and 1.0 g NO_x per vehicle mile. Successive deferrals of deadlines, and relaxation of emission limitations that have resulted from hearings before the EPA, already have delayed such deadlines. A federal emission-testing procedure assures that every vehicle conforms to the standards. A 78% reduction in hydrocarbons and 60% reduction in carbon monoxide occurred by 1985 over 1960, despite a 52% growth in the total numbers of vehicles during this period. Reductions in oxides of nitrogen and the resulting ozone were less encouraging. In fact, trend analysis of ozone, oxides of nitrogen and hydrocarbons in the South Coast Air Basin of California shows only a very weak downward trend between 1968 and 1984. As might be expected, there was considerable variation from year to year depending on climatic conditions. The ozone variation was explained largely by the meteorological variable (Kuntasal & Chang, 1987).

What would be the results of a cost–benefit analysis if one compared the effectiveness of control procedures with their costs? After all, when one adds the costs of control equipment and the costs of operating regulatory agencies over just the past few decades, the sum reaches billions of dollars. What have we to show for it? How close have we come to achieving the Ambient Air Quality Standards promulgated in the 1970s (Table 16.1)? With industrial pollutants, including sulfur dioxide, fluorides and lead, the gains have been considerable, and no longer do these pose a problem. With ozone and largely automobile-related pollutants, on the other hand, there have been no measurable reductions in concentrations during the past decade of monitoring. Neither are improvements apparent for oxides of nitrogen, hydrocarbons or carbon monoxide.

The responsibility for air quality standards in Canada is left to the provinces. Air quality regulations, which are meant to be legally enforceable, apply in Alberta, Saskatchewan, Quebec and New Brunswick. The values cited, as well as for the 'criteria' and 'objective' values that provide recommended goals in the other provinces, vary considerably. They are

Table 16.1. National US primary and secondary ambient air quality standards[1]

Pollutant	Type of standard	Averaging time	Frequency parameter	Concentration $\mu g/m^3$	ppm
SO_2	Primary	24 h	Annual maximum[2]	365	0.14
	Secondary	3 h	Annual maximum[2]	1 300	0.5
CO	Primary	1 h	Annual maximum[2]	40 000	35
	Secondary	8 h	Annual maximum[2]	10 000	9
O_3	Primary and secondary	1 h	Annual maximum[2]	235	0.12
NO_2	Primary and secondary	1 yr	Arithmetic mean	100	0.05

[1] Adapted from *Air Pollution Control Association Directory and Resource Book* (1981–1982), APCA, Pittsburgh.
[2] Annual maxima are not to be exceeded more than once per year.

essentially in the range of the federal objectives. Three categories are noted having different values, as shown in Table 16.2. The *desirable* range defines the long-term goal for air quality; the *acceptable* range is intended to provide adequate protection against adverse effects on soil, water, vegetation, materials, animals, visibility, personal comfort and well-being. The *maximum tolerable* level denotes a pollutant concentration that requires abatement without delay due largely to substantial risks to public health. But whatever the approach, the difficulty lies in attainment.

Air quality in countries other than in North America

In Europe the air quality standards approach is used largely to supplement the emission control approach; and some standards have been set.

However, there is no consistency in the air quality standards among the countries of Europe or between Europe and North America, even for sulfur dioxide, which has been studied for the longest period. Standards for a 30-minute period range from 0.07 mg/m³ (0.028 ppm) (zone I) in protected areas of Austria to 0.75 mg/m³ (0.30 ppm) in Italy. Recommendations over a 24-hour period range from 0.07–0.10 mg/m³ in Austria to 0.38 mg/m³ (0.15 ppm) in Israel and Italy. Standards of both Western and Eastern European countries fall within this range. Air quality standards for these and several other countries around the world are summarized in Table 16.3.

Table 16.2. Air quality objectives in Canada[1]

Pollutant	Federal objective ($\mu g/m^3$)		
	Desirable range	Acceptable range	Tolerable range
Sulfur dioxide			
1-h average	0–450	450–900	–
24-h average	0–150	150–300	300–800
Annual arithmetic mean	0–30	30– 60	–
Nitrogen dioxide			
1-h average	–	0–400	400–1000
24-h average	–	0–200	–
Annual arithmetic Mean	0–60	60–100	–
Oxidants (ozone)			
1-h average	0–100	100–160	160–300
24-h average	0–30	30–50	–
Annual arithmetic Mean	–	0–30	–

[1] Adapted from *Air Pollution Control Association Directory and Resource Book* (1981–1982), APCA, Pittsburgh.

While sulfur oxides are well addressed, along with suspended particulates and nitrogen oxides, ozone is not directly addressed except in North America.

In an effort to provide some uniformity, the World Health Organization (WHO) Regional Office for Europe submitted air quality guidelines for major urban air pollutants in 1985 (World Health Organization, 1987). WHO recognized the desirability of taking an integrated view of both health and ecological effects in air quality management, and considered the ecological effects of sulfur oxides, nitrogen oxides and ozone/photochemical oxidants on terrestrial vegetation. The guideline values are intended to provide just that: to provide recommendations for effective protection against recognized hazards. While they provide general recommendations that would hopefully facilitate the standard-setting process, there is no legal basis. Based on a thorough review of European and US literature, WHO concluded that the lowest sulfur dioxide concentrations at long-term exposures capable of reducing plant growth were in the range between 40 and 80 $\mu g/m^3$ (15–30 ppb). A maximum annual mean sulfur dioxide concentration of 30 $\mu g/m^3$ (12 ppb) was recommended. Similar studies of the ozone literature indicated that the lowest harmful ozone concentrations during the growing season were 300 $\mu g/m^3$ for 1 hour (150 ppb), 65 $\mu g/m^3$ (32 ppb) for 24 hours and 60 $\mu g/m^3$ for a growing season of 100 days (30 ppb). Mean ozone concentration should stay below 200 $\mu g/m^3$ (102 ppb) for 1 hour, 65 $\mu g/m^3$

Table 16.3. Air quality standards for selected pollutants in several countries around the world[1]

Country	Nitrogen oxides (mg/m^3)	Sulfur oxides (mg/m^3)
Austria	None	Zone I 0.07–0.15/30 min 0.07–0.10/24 h Zone II 0.20–0.30/30 min 0.30–0.30/24 h
Bulgaria	0.085/30 min 0.085/24 h	0.5/30 min (SO_2) 0.05/24 h
FRG	0.08/30 min (NO_2) 0.2/30 min (NO)	0.14/y
Finland	0.5/1 h (NO_2)	0.300/24 h 0.07/yr
Israel	1.0/30 min	0.75/30 min 0.26/24 h
Italy	0.6/30 min (NO_2) 0.2/24 h	0.75/30 min 0.38/24 h
Japan	0.075–0.1/24 h	0.26/1 h 0.10/24 h
Norway	0.4/1 h 0.2/24 h	0.4/1 h 0.2/24 h
USSR	0.085/min (NO_2) 0.085/24 h	0.50/30 min 0.05/24 h

[1] Adapted from *Air Pollution Control Association Directory and Resource Book* (1981–1982), APCA, Pittsburgh.

for 24 hours and 60 μg/m^3 for the growing season if all plants are to be protected.

European countries share the same broad air shed and encounter rather similar air pollution problems. Inasmuch as the sources are often similar and pollutants do not stop at borders, this is not surprising. Yet, the control approaches have not been uniform.

In Germany, a 1964 'Joint Ministerial Bulletin' prescribed control devices and pollutant emission limits for all industrial sources. In 1978 the International Union of Forestry Research Organizations (IUFRO) proposed a resolution about the necessity to limit emissions of sulfur dioxide to 0.05 mg/m^3 (19 ppb) and hydrogen fluoride to 0.3 μg/m^3 (0.4 ppb) to protect the health of the spruce forests.

A national law in France, passed in 1961, was an enabling action to prevent air pollution and noxious odors from domestic or industrial sources.

In case of non-compliance the operation may be temporarily suspended.

An Italian law of 1966 provides the general framework for control of air pollution. The law covers all fuel-burning facilities, including industrial and residential as well as vehicle emissions. A Commission is charged with examining all aspects of air pollution and promoting research in the field.

Scandinavian legislation also requires a permit to spread any substances that might cause harm or inconvenience to persons over a large area.

The only real solution, at least to the photochemical pollution associated with automobiles, is to minimize their use. The Principality of Liechtenstein addressed this in 1987 by offering free mass transit to and from the capital and ten other villages in Liechtenstein for a trial period of one year. This decision was activated largely by the recognition that pollution was taking its toll of trees.

Much of the early Western European air pollution legislation is being superseded by regulations passed by the Commission on European Community members. One important achievement in the 1980s was the setting of compulsory air quality standards to be respected by all the member states. Standards were set for sulfur dioxide, suspended particulates matter, lead and nitrogen dioxide.

The 1976 Act to Prevent Air Pollution in Czechoslovakia is representative of Eastern Europe. Limits were set on the discharge of fly ash, sulfur dioxide and specified harmful substances from any source. Air pollution emissions in excess of these limitations are subject to a fine. Since it is often more expedient to pay the fine than control the pollutant, this is the route often taken.

SUGGESTED READING

Covello, V.T., ed. (1985). *Environmental Impact Assessment, Technology Assessment, and Risk Analysis: Contributions from the Psychological and Decision Sciences.* Springer-Verlag, New York, NATO Advanced Science Institutes Series G, No. 4. x + 1069 pp. illus. (From an institute in Bourgy-St.-Maurice, France, August 1983).

Rowe, W.D. (1983). In *Evaluation Methods for Environmental Standards.* (F.J. Hageman, ed.) CRC Press, Boca Raton, FL. xxii + 282 pp., illus.

Stern, A.C., ed. (1985). *Supplement to Management of Air Quality*, Vol. 8. Academic Press, Orlando, FL. xii + 206 pp., illus.

CHAPTER 17

Prospects

Air pollution biology has progressed enormously from the early days when injury was measured by the severity of visible symptoms. Pollution control technologies now have advanced to a degree whereby visible symptoms rarely occur when such technology is utilized. On the other hand, we have learned that it is the subliminal, not visible, 'symptoms' that are most critical to plant health and production. Much is known about the pollutant concentrations that have adverse effects, and we know something of the effects these effects have on production and their costs to society. Yet, much remains to be discovered. Our data base is still inadequate, especially in non-urban areas; dose-response relations are poorly understood, pollutant–pathogen interactions, mycorrhizal and environmental interactions have received only preliminary attention, and genetic mechanisms largely remain unstudied. Long-term effects on natural ecosystems and losses of biodiversity have received almost no attention. All of these questions must be answered in order to provide a basis to establish meaningful and justifiable air quality standards.

EXPOSURE CHARACTER TRENDS

An extensive network of monitoring stations in the USA and Europe centers around the major urban areas. But vast gaps remain over most of the non-urban rangelands and forests, which may be subjected to higher ozone concentrations due to long-range transport and relative absence of scavenging molecules. Available data provide only general, not specific, information. Across most of the USA, the average regional 7-hour growing season ozone concentrations are about 50 ppb—roughly twice what might be considered background. Seasonal concentrations are relatively uniform except for local stations and specific areas, such as downwind from the Los Angeles Basin, where mean concentrations are closer to 70 ppb (National Acid Precipitation Assessment Program, 1986). Peak concentrations, and numbers of hours

222

exceeding a specific concentration, values that probably are more relevant to plant responses, are not routinely reported. Thus it remains difficult to relate effects to the appropriate exposure dynamics.

Annual average concentrations of sulfur dioxide and oxides of nitrogen in urban environments are at least ten times greater than in rural areas, due to the abundance of industry and people in such areas.

One obvious purpose in monitoring is to evaluate the success of control programs by learning if pollutant concentrations are diminishing. Unfortunately, a valid network has not been in operation long enough to provide meaningful trend data. Reliable ozone data only extend back to 1978, and the wet deposition network was completed only in 1985, although 19 National Acid Deposition sites have been operating in the USA since 1978.

Available data provide some valuable insights. Since reaching a peak in 1973, sulfur dioxide emissions in the USA have decreased by about 28%. In the northeast, emissions have dropped by 19% despite an increase in coal consumption. Annual sulfur dioxide levels measured at 302 sites decreased 37% from 1977 to 1986. Sulfur dioxide emissions in Europe also have decreased during this period.

The gains may not be due solely to emission control efforts. It was during this same period after 1974, that oil prices increased greatly. Coinciding with the price rises and attending inflation, energy conservation was practiced more than before, which resulted in what is essentially a moratorium on building new power plants, which continues today. This, as much as regulatory effectiveness, may have accounted for much of the sulfur dioxide reductions. Nitrogen oxide emissions increased steadily to about 1977, after which they have remained fairly constant, again, perhaps, coinciding with energy conservation at least in the USA. Ammonia has added significantly to the nitrogen loading of the environment, yet data regarding exposure dynamics are sparse. Eutrophication of the environment very well may be more important to plant health than acidification (Heil *et al.*, 1988).

The US EPA bases trends in ozone concentrations on the second highest daily maximum 1-hour values. Based on data from 242 monitoring sites in the USA, ozone concentrations decreased 21% between 1977 and 1986. This amounts to a decrease from 15 to 12 ppb. However, much of the difference appears to have been due to a change in calibration methods and instrumentation in 1978. It is at least as significant to note that oxides of nitrogen and VOCs (volatile organic compounds), which go into producing ozone, remained essentially constant throughout this period. The 5-year period of 1982–1986 showed a 4% decrease based on data from 539 sites. But the 1983 values were clearly higher than the other four years, emphasizing the important influence of meteorological conditions in being more conducive to ozone formation in certain years. Some cities showed considerable increases in ozone concentrations during this period. Boston, for instance, showed a 33% increase, but this value was driven by very low values in 1981. The statistical significance of a nation-wide 5% decrease is debatable, particularly for those living in areas where concentrations continue to climb. In short, trend analyses do not suggest that a meaningful decrease in ozone concentration is likely for many years. During years of unfavorable meteorological conditions, such as 1983, they may well increase. Exceptionally high ozone concentrations in 1988 were attributed to record high summer temperatures. There is no question that increases consistent with climatic patterns will appear in specific geographical areas.

Additional research is needed to determine the most valid exposure regimes to use for dose–response studies. But how is it possible to simulate the ever-changing concentrations and exposure periods found in ambient air environments? This infinite array of choices makes it desirable to find a realistic surrogate exposure regime. A single value may not be possible since nitrogen dioxide and ozone concentrations, as well as the time of peak concentrations, vary from one city to another, from the strong midday peaks in Los Angeles to two peaks in El Paso and no well-defined peak in Chicago. Seasonal peaks also vary. How might this influence different stages of plant growth? And how significant are fall and winter exposures to damaging perennial plants? Perhaps more significantly, how do multiple exposures over a period of years influence these species?

Co-occurrences of pollutants are well documented, but their effects remain controversial. Sulfur dioxide may now be minimal in many countries, but concentrations in Europe remain sufficiently high that interactions with ozone and other pollutants are entirely possible, if not likely. Oxides of nitrogen and ammonia are more prevalent than ever. And what about hydrogen peroxide? This strong oxidizer is present at higher elevations at what may be significant concentrations, but it is scarcely being monitored or researched. Presence of peroxyacetyl nitrates has been substantiated and studied since the 1960s, but technical difficulties have discouraged research to monitor its presence or interactions with other pollutants. A greater monitoring data base for these and other pollutants is needed. The same holds true for pollutant combinations involving acid deposition. Studies to date have shown little interaction between these in terms of effects, but exposures have been for rather short duration. Less is known regarding long-term interactions of these combinations.

BIOLOGICAL RESEARCH NEEDS

Agricultural production

Despite tremendous gains in knowledge through the NCLAN program and other research, many unanswered questions remain regarding ozone impacts.

In order to set meaningful air quality standards we should know more. Specifically we should know the concentrations and durations of exposure to which different plant species respond. We have reasonably good knowledge of how 7- and 12-hour exposures affect production of major crops. While this provides a general simulation of real world conditions, it does not fully simulate the many variations in exposure dynamics to which plants may be exposed. The average and peak concentrations primarily studied provide only the most general surrogate for accurately assessing what happens in the field. The number of hours above a given concentration, a value that may be far more meaningful, has only begun to be studied. Even more neglected has been the significance of the time of day during which plants are exposed to a pollutant. The middle of the day is clearly a reasonable place to start, but in many non-urban areas the periods of highest concentrations are in the late afternoon, evening or even late at night. How do plants respond at these times?

Ecosystem and plant community responses

Ecosystems and plant communities involve interactions between a myriad of diverse species. Some species are sensitive to a given air pollutant; some are not; and most lie somewhere in between. We know how only a few of these plants respond to pollutants and virtually nothing of how they respond collectively as an ecological system.

Studies of natural systems have been limited to short-term exposures, often of only a few hours duration. This has some relevance to annual species, but virtually none to perennial plants. Natural systems are composed of many perennial herbaceous and woody species. How do these respond when exposed to even low pollutant concentrations year after year? And what will be the long-term impact on the diversity of the system?

And what about secondary impacts? Sustained inputs of sulfur and nitrogen compounds are bound to modify the character of soils. Acidification is only one such modifier. The added nutrient level in itself, especially as it might influence broader nutrient balance, or imbalance, may be far more significant in the long run.

The rhizosphere is a vital component of any ecosystem, yet very little is known regarding how pollutants influence it. Similarly, the mycorrhizae may be affected directly or indirectly; these may be among the most sensitive components of all to pollutant stress. The effect most likely would be indirect, as when carbon allocation is disrupted or root growth becomes excessive through abnormal nitrogen input. Effects could also be direct where the fungal component is especially sensitive to a pollutant. But this is less likely since pollutants would be filtered by the soil particles or dissolved in the soil solution and would not readily enter the soil in significant amounts.

Plant Stress in Air Pollution

The direct or indirect effects of pollutants on these systems may, over time, by favoring one species over another, lead to the disappearance of the more sensitive species, creating a loss of diversity in the ecosystem. This has clearly been brought about in the most devastated areas, but we have no answers in areas where visible symptoms and direct mortality are absent.

Forests are the most abundant plant communities in many countries. They are clearly the most important in terms of timber production and recreation. Species such as pines are among the most sensitive plants to ozone. Despite exposures to often high concentrations, these are not always the species sustaining decline. Very little is known regarding the sensitivity of most forest species or the exposure character that might be most critical. Although ozone has been responsible for decline in the USA, other pathogens, or at least interactions, are more reasonably implicated in Europe.

Considerable variation in sensitivity exists among individual trees, apparently due to differences in their genetic character. But what is the basis for such variability? What type of inheritance is involved, and what might be the long-term implications and influence on community structure and biodiversity?

Interactions

Pollutant–pathogen interactions, pollutant–pollutant interactions and pollutant–environment interactions—all of these are known to seriously modify the response of plants to air pollutants. The abiotic environment appears to have the strongest influence. While nutrient, soil and temperature relations are important, moisture is most critical. Other factors, such as CO_2 concentrations, may be at least equally significant, but they have received scant attention. Agricultural chemicals and crop management practices fall into this category of warranting study.

INDUSTRIALIZING COUNTRIES

Are the rapidly industrializing, or developing, countries destined to repeat the ecological tragedies begun by the industrialized nations a century ago? They seem to be! Many nations are tempted to equate industry with affluence. In the last decade a number of countries have experienced dramatic industrial growth. The nations include some of Europe's poorer countries, as well as Mexico, South Korea, Taiwan, Singapore, India, and several countries in South America, Africa and Asia. Air pollutant problems are exacerbated by overpopulation, enormous volumes of sewage and solid wastes and critical water pollution. Many of the tropical and subtropical emerging nations are trading their rainforests for relative wealth, from lumber and charcoal and land to graze cattle, with devastating ecosystem effects, and perhaps even with global climatic consequences.

Some nations have encouraged multinational corporations to locate within their borders; others have purchased sometimes outdated technology or even whole industrial plants. Economic opportunities have increased often at the expense of the environment. Larger cities, such as Mexico City, Sao Paulo, Seoul, Jakarta, Lima and Calcutta, show the most common examples of serious environmental contamination (Leonard, 1985). Many smelters, with coal-fired power plants to provide the electricity, cement plants and petro-chemical plants, are all being constructed with minimum concern for the pollutants emitted. Facilities may be new, but in some cases, as in Cubatao, Brazil, they are purchased from industrial nations whose environmental laws no longer permit their operation, lacking pollutant control technology. Many important plant species, including *Podocarpus*, *Cassia*, *Jacaranda* and *Ilex*, are no longer found near Cubatao, and loss of soil stability has precipitated numerous damaging land flows.

Tropical forests near an aluminum reduction plant in Minas Gerais, Brazil, have been seriously affected by a serious loss in biodiversity as sensitive species are eliminated. Old industries, built before current control technology was widely used, present some of the greatest problems since regulations are generally less stringent.

In the Goktas area of western Turkey, copper has been refined only since 1951. The smelter was established at the bottom of a narrow valley, with concern only for proximity to the ore. Apparently there was no regard for the pollutant-sensitive red pine (*Pinus brutia*) forest occupying the surrounding slopes. The classic decline that ensued followed the patterns known in Europe and North America almost a century earlier. Drastic soil erosion involved over 7800 hectares (about 19 300 acres). A sulfur recovery system was installed in 1965, but it was already a case of too little, too late (Cepel, 1988).

Also in Turkey, the Yatagan thermal power station has been operational since 1982. The sulfur dioxide released caused serious injury to 40 000 hectares (98 800 acres) of pine (*P. brutia*) forests in the area.

Industrial expansion also takes its toll in agricultural land. The land itself is used, but beyond this heavy metal accumulation can be dangerous to human health when crops such as rice (*Oryza sativa*) are grown on polluted lands (Ray, 1988). Crops near several cities in India suffer huge crop losses due to a mixture of gaseous pollutants, most notably sulfur dioxide and oxides of nitrogen (Pandya & Bedi, 1988). Yield losses approach 95% near the sources.

In many nations private industry is forced by law to adhere to strict environmental laws. Costs are borne by industry. National budgets, especially in developing nations, often cannot afford the costs of a pollutant-free environment. It is glib to say that they cannot afford a polluted environment. Economic balances must be a consideration. At the very least, though, such balances should be calculated and the costs of the pollution figured in the cost–benefit equations.

ATMOSPHERIC ENERGY BALANCE

Perhaps the most serious future threat from air pollution is the global rise in several 'natural' and unnatural atmospheric substances that are the result of industrial processes—the burning of fossil fuels (coal, oil, gas). These substances include carbon dioxide, methane, other hydrocarbons and numerous anthropogenic chemicals including CFC's. Although some are 'natural' atmospheric components, their rising concentrations are not natural. Climatic models predict rising global temperatures of some 1–5 °C by the middle of the next century, accompanied by such planetary disasters as the melting of the polar ice-caps, a 5–6 m rise in sea level with attendant coastal flooding, drastic shifts in the world's precipitation patterns, and attendant changes in the earth's vegetation zones and agricultural patterns. All of these are secondary effects of air pollution, but they are overwhelmingly important and should be included among the traditional approaches to the study of air pollution effects on plants. Ultimately, these global effects may exceed in economic impact all the direct effects of all other 'classic' air pollutants combined. The magnitude of these predicted global effects is an example of how careless humans have been in their concern for their own environment.

SUGGESTED READING

Briggs, D.J., & Courtney, F.M. (1987). *Agriculture and Environment: The Physical Geography of Temperate Agricultural System.* Longman, New York. xiv + 442 pp., illus.

Goudie, A. (1982). *The Human Impact: Man's Role in Environmental Change.* MIT Press, Cambridge, MA. xii + 316 pp., illus.

Kates, R.W., Ausubel, J.H., & Berberian, M., eds (1985). *Climatic Impact Assessment: Studies of the Interaction of Climate and Society.* SCOPE, International Council of Scientific Unions. Wiley, New York. SCOPE 27. xxiv + 625 pp.

Knoepfel, H. (1986). *Energy 2000: An Overview of the World's Energy Resources in the Decades to Come.* Gordon & Breach, New York. viii + 181 pp., illus.

Prance, G.T., ed. (1986). *Tropical Rain Forests and World Atmosphere.* Westview press, Boulder, Colorado. xxii + 105 pp., illus. (AAAS Symposium. 101, New York, May 1984).

Rosenblum, J.W. (1983). *Agriculture in the Twenty-First Century.* Wiley–Interscience, New York. xxii + 415 pp., illus. From a symposium at Richmond, VA, 1983.

Tickell, C., ed. (1986). *Climatic Change and World Affairs.* Center for International Affairs, Harvard University, and University Press of America, Lanham, MD. xvi + 76 pp., illus.

References

Abeles, F.B. (1973). *Ethylene in Plant Biology*, 392 pp., Academic Press, New York.

Abrahamsen, G. (1980). Acid precipitation, plant nutrients, and forest growth. In D. Drablos & A. Tollan (eds), *Proc. Int. Conf. Ecol. Impacts Acid Precipitation*, pp. 58–63, SNSF Project, Ås, Norway.

Abrahamsen, G. (1984). Effects of acid deposition on forest soil and vegetation. *Phil. Trans. R. Soc. London*, **305**, 369–372.

Adamait, E.J., Ensing, J., & Hofstra, G. (1987). A dose–response function for the impact of O_3 on Ontario-grown white bean and an estimate of economic loss. *Can. J. Bot. Sci.*, **67**, 131–136.

Adams, D.F., & Sulzback, C.W. (1961). Nitrogen deficiency and fluoride susceptibility of bean seedlings. *Science*, **133**, 1425–1426.

Adams, D.F., Mayhew, D.J., Gnagy, R.M., Richey, E.P., Koppe, R.K., & Allen, I.W. (1952). Atmospheric pollution in the ponderosa pine blight area. *Ind. Eng. Chem.*, **44**, 1356–1365.

Adams, R.M., Hamilton, S.A., & McCarl, B.A. (1984). Economic effects of ozone on agriculture. Report no. EPA 600-3-84-090. US Environmental Protection Agency, EPA Research Lab., Corvallis, OR.

Adams, R.M., Crocker, T.D., & Katz, R.W. (1984a). Assessing the adequacy of natural science information: a Bayesian approach. *Rev. Econ. Stat.* **66**, 568–575.

Adams, R.M., Hamilton, S.A., & McCarl, B.A. (1985). An assessment of the economic effects of ozone on U.S. agriculture. *J. Air Poll. Contr. Assoc.* **35**, 938–943.

Adams, R.M., Hamilton, S.A., & McCarl, B.A. (1986). The benefit of pollution control: The case of ozone and U.S. agriculture. *Am. J. Agric. Econ.*, **68**: 886–893.

Adams, R.M., Glyer, J.D., & McCarl, B.A. (1989). The NCLAN economics assessment: Approach, findings and implications. EPA Public (in press).

Adelpipe, N.O., Hofstra, G., & Ormrod, D.P. (1972). Effects of sulfur nutrition on phytotoxicity and growth responses of bean plants to ozone. *Can. J. Bot.* **50**, 1789–1793.

Adelpipe, N.O., Khatamian, H., & Ormrod, D.P. (1973). Stomatal regulation of ozone phytotoxicity in tomato. *Z. Pflanzenphysiol.* **68**, 323–328.

Adelpipe, N.O., & Ormrod, D.P. (1974). Ozone-induced growth suppression in radish plants in relation to pre- and post-fumigation temperatures. *Z. Pflanzen Physiol.* **71**, 281–287.

Ahmadjian, V. (1967). *The Lichen Symbiosis*, 152 pp., Blaisdell, Waltham, MA.

Ahmadjian, V. (1974). 'Lichen'. *Encyclopaedia Britannica*, **228**, 74–80.

Air Pollution Control District (1951). County of Los Angeles 1950–1951. *2nd Tech. and Admin. Rept.*. 51 pp.

Alscher, R. (1984). Effects of SO_2 on light-modulated enzyme reactions. In M.J. Kozial & F.R. Whatley (eds), *Gaseous Air Pollutants and Plant Metabolism*, pp. 181–200, Butterworths, London.

Altshuller, A.P. (1986). Review paper: The role of nitrogen oxides in non urban ozone formation in the planetary boundary layer over North America, north Europe and adjacent areas of ocean. *Atmos. Environ.*, **20**, 245–265.

Altshuller, A.P. (1987). Estimation of natural background of ozone present at surface rural locations. *J. Air Poll. Contr. Assoc.*, **37**, 1409–1417.

Amundson, R.G., & McLean, D.C. (1982). Influence of oxides of nitrogen on crop growth and yield: An overview. In T. Schneider & C. Grant (eds), *Air Pollution by Nitrogen Oxides*, pp. 501–510, Elsevier, Amsterdam.

Amundson, R.G., & Weinstein, L.H. (1980). The effect of SO_2 and NO_2 alone and in combination on the yield of soybean. *Plant Physiol.* **65** (suppl.), 152.

Amundson, R.G., & Weinstein L.H. (1981). Joint action of sulfur dioxide and nitrogen dioxide on foliar injury and stomatal behavior in soybean. *J. Environ. Qual.*, **10**, 204–206.

Anderson, F.K. (1966). *Air Pollution Damage to Vegetation in Georgetown Canyon, Idaho*. MSc thesis, University of Utah, Salt Lake City, 102 pp.

Anderson, F.K., & Treshow, M. (1984). Responses of lichens to atmospheric pollution. In M. Treshow (ed.), *Air Pollution and Plant Life*, pp. 259–289, Wiley, Chichester.

Anderson, H.W. (1931). Problems on spray to the peach. *Trans. Sec. State Hort. Soc.* **65**, 454–465.

Anderson, R.L., Brown, H.O., Chevone, B.I., & McCartney, T.C. (1988). Occurrence of air pollution symptoms in eastern white pine in the southern Appalachian Mountains. *Plant Dis.*, **72**, 130–132.

Arndt, U., Seufert, G., & Nobel, W. (1982). Die Beteiligung Von Ozon an der Komplex Krankheit der Tanne (*Abies alba*) Millteine prüfenswerte Hypothese. Staub-Reinholt. *Luft*, **42**, 243–247.

Ashenden, T.W. (1979a). The effects of long term exposures to SO_2 and NO_2 pollution on the growth of *Dactylis glomerata* L. and *Poa pratensis* L. *Environ. Pollut.*, **18**, 249–258.

Ashenden, T.W. (1979b). Effects of SO_2 and NO_2 pollution on transpiration in *Phaseolous vulgaris* L. *Environ. Pollut.*, **18**, 45–50.

Ashenden, T.W., & Williams, I.A.D. (1980). Growth reduction in *Lolium multiflora* and *Phleum pratense* as a result of SO_2 and NO_2 pollution. *Environ. Pollut.* **21**, 131–139.

Asher, J.E. (1956). Observations and theory on 'X' disease or needle dieback. File report, Arrowhead Ranger District. San Bernadino National Forest, California.

Ashmore, M.R., Bell, J.N.B., & Reily, C.L.A. (1978). A survey of ozone levels in the British Isles using indicator plants. *Nature (London)*, **276**, 813–815.

Ashmore, M., Bell, N., & Rutter, J. (1985). The role of forest damage in West Germany. *Ambio*, **14**, 81–87.

Ballantyne, D.J. (1972). Fluoride inhibition of the Hill reaction in bean chloroplasts. *Atmos. Environ.*, **6**, 267–273.

Barkman, J.J., Rose, F., & Westhoff, V. (1969). The effects of air pollution on non-vascular plants. In *Air Poll. Proc. Eur. Congr. Influence Air Pollut. Plants*, pp. 237–241.

Bauch, J. (1983). Biological alterations in the stem and root of fir and spruce due to pollution influence. In B. Ulrich & J. Pankrath (eds), *Effects of Accumulation of Air Pollutants in Forest Ecosystems*, pp. 377–386, Reidel, Dordrecht.

Baumback, G. (1986). Occurrence of gaseous pollutants in forest stands. In H.W. Georgii (ed.), *Atmospheric Pollutants in Forest Areas*, pp. 177–187, Reidel, Dordrecht.

Becker, K.H., Fricke, W., Lobel, J., & Schurath, U. (1985). Formation, transport, and control of photochemical oxidants. In R. Guderian (ed.), *Air Pollution by Photochemical Oxidants*, pp. 1–125, Springer-Verlag, Berlin.

Beckerson, D.W., & Hofstra, G. (1979). Response of leaf diffusion resistance of radish, cucumber and soybean to O_3 and SO_2 singly or in combination. *Atmos. Environ.*, **13**, 1263–1268.

Beckerson, D.W., & Hofstra, G. (1979). Stomatal response of white bean to O_3 and SO_2 singly or in combination. *Atmos. Environ.*, **13**, 533–535.

Bell, J.N.B. (1984). Air pollution problems in Western Europe. In M.J. Kozial & F.R. Whatley (eds), *Gaseous Air Pollutants and Plant Metabolism*, pp. 3–24, Butterworths, London.

Bell, J.N.B., & Cox, R.A. (1975). Atmospheric ozone and plant damage in the United Kingdom. *Environ. Pollut.*, **8**, 163–170.

Bell, J.N.B., Rutter, A.J., & Relton, J. (1979). Studies on the effects of low levels of sulphur dioxide on the growth of *Lolium perenne* L. *New Phytol.*, **83**, 627–643.

Benedict, H.M., Ross, J.M., & Wade, R.W. (1964). The disposition of atmospheric fluorides by vegetation. *Int. J. Air Water Pollut.* **8**, 279–289.

Benedict, H.M., Miller, C.J., & Smith, J.S. (1973). *Assessment of Economic Impact of Air Pollutants on Vegetation in the United States: 1969–1971*. EPA 650/5-78-002, Stanford Research Institute, Menlo Park, CA.

Bennett, J.H. (1975). Acute effects of combinations of sulfur dioxide and nitrogen dioxide on plants. Environ. Pollut., **9**, 127–132.

Bennett, J.H., Hill, A.C., & Gates D.M. (1973). A model for gaseous pollutant sorption by leaves. *J. Air Poll. Contr. Assoc.*, **23**, 957–962.

Bennett, J.H., Lee, E.H., & Heggstad, H.E. (1984). Biochemical aspects of plant tolerance to ozone and oxy radicals; superoxide dismutase. In M.J. Kozial & F.R. Whatley (eds), *Gaseous Air Pollutant and Plant Metabolism*, pp. 413–424. Butterworths, London.

Bennett, J.P., Anderson, R.L., Campana, R., Clarke, B.B., Houston, D.B., Linzon, S.N., Mielke, M.E., & Tingey, D.T. (1986). Needle tip necrosis on eastern white pine in Acadia National Park. Maine Workshop to determine possible causes. Air Quality Division, NPS, Denver.

Benoit, L.F., Skelly, J.M., Moore, L.D., & Dochinger, L.S. (1985). Radial growth reductions of *Pinus strobus* L. correlated with foliar ozone sensitivity as an indicator of ozone-induced losses in eastern forests. *Can. J. For. Res.*, **12**, 673–678.

Benson, E.J., Krupa, S., Tengy, P.S., & Welsch, P.E. (1982). *Economic Assessment of Air Pollution Damages to Agriculture and Silvicultural Crops in Minnesota*. Final Report to Minnesota Pollution Control Agency.

Benson, N.R. (1959). Fluoride injury on soft suture and splitting of peaches. *Proc. Am. Soc. Hort. Sci.*, **74**, 184–198.

Berry, C.R. (1961). White pine emergence tipburn, a physiogenic disturbance. *US Forest Service Southeast Forest Exp. Sta., Sta. Papers*, **130**, 1–8.

Berry, C.R., & Ripperton, L.A. (1963). Ozone, a possible cause of white pine emergence tipburn. *Phytopathology*, **53**, 552–557.

Berry, C.R., & Hepting, G.H. (1964). Injury to eastern white pine by unidentified atmospheric constituents. *Forest Sci.*, **10**, 2–13.

Betzer, P.R., Byrne, R.H., Acker, J.G., Lewis, C.S., & Jolley, R.R. (1984). The oceanic carbonate system: A reassessment of biogenic controls. *Science*, **226**, 1074–1077.

Bingham, F.T., Page, A.L., & Strong, J.E. (1980). Yield and Cd content of rice grain in relation to addition rate of Cd, Cu, Ni, Zn and liming. *Soil Sci.*, **130**, 32–38.

Bisessar, S. (1982). Effect of ozone antioxidant protection and early blight on potato in the field. *J. Am. Soc. Hort. Sci.*, **107**, 597–599.

Bisessar, S., & Palmer, K. (1984). Ozone, antioxidant smog, and *Meloidogyne hapla* effects on tobacco. *Atmos. Environ.*, **18**, 1025–1027.

Bjorkman, E. (1949). Ecological significance of ecototrophic mycorrhizae associations of forest trees. *Sven. Bot. Tidskr.*, **42**, 1–223.

Black, V.J. (1982). Effects of sulfur dioxide on physiological processes in plants. In M.H. Unsworth & N.P. Ormrod (eds), *Effects of Gaseous Air Pollution in Agriculture and Horticulture* pp. 225–246, Butterworths, London.

Black, V.J. (1984). Effects of sulphur dioxide on physiological processes in plants. In M.H. Unsworth, & D.P. Ormrod (eds), *Effects of Gaseous Air Pollution in Agriculture and Horticulture*, pp. 67–92, Butterworths, London.

Black, V.J., & Unsworth, M.H. (1979). Effects of low concentrations of sulphur dioxide on gas exchange of plants and dark respiration of *Vicia faba*. *J. Exp. Bot.*, **30**, 473–483.

Bleasdale, J.K. (1952). *Atmospheric Pollution and Plant Growth.* PhD thesis, University of Manchester.

Bleasdale, J.K.A. (1973). Effects of coal-smoke pollution gases on the growth of Ryegrass (*Lolium perenne* L.). *Environ. Pollut.*, **5**, 275–278.

Bobrov, R.A. (1952). The effect of smog on the anatomy of oat leaves. *Phytopathology*, **42**, 558–563.

Bolay, A., & Bovay, E. (1965). Observations sur les dégâts provoques par les composes flúores en valais. *Agr. Romande*, **4**, 43–46.

Bolin, B., Doos, B.R., Jager, J., & Warrick, R.A. (eds) (1986). *The Greenhouse Effect, Climate Change and Ecosystems.* International conference at Villach, Austria, October 1985. Wiley, New York.

Bordeau, P., & Treshow, M. (1978). Ecosystems response to pollution. In G.C. Butler (ed.), *Principles of Ecotoxicology*, pp. 313–322, Wiley, Chichester.

Bosch, Von C., Pfannkuch, E., Baum, U., & Rehfuess, K.E. (1983). Über die Erkrankung der Fichte (*Picea abies* Karst.) in den Hochlagen des Bayerischen Waldes. *Forstw. Cbl.* **102**, 167–181.

Boyle, G.W., Jr, & Jonasson, L.R. (1973). The geochemistry of arsenic and its use as an indicator element in geochemical prospecting. *J. Geochem. Explor.*, 2, 251–296.

Bradley, C.E., & Haagen-Smit, A.J. (1950). Application of rubber in the quantitative determination of ozone. Report to the Los Angeles County Air Pollution Control District, Nov. 10, 1950.

Bradley, C.E., & Haagen-Smit, A.J. (1951). The application of rubber in the quantitative determination of ozone. *Rubber Chem. Technol.*, **24**, 750.

Braekke, F.H. (1979). Boron deficiency in forest plantations on peatland in Norway. *Meddr. Norsk. Inst. Skogforsk.*, **35**, 213–236.

Braekke, F.H. (1979). Boron deficiency on afforested peatland in Norway. In *Proc 6th Internat. Peat Congress*, Diluth, MA, Aug. 17–23, pp. 369–375.

Brechtel, H.M., Balazs, A., & Lehnardt, F. (1986). Precipitation input of inorganic chemicals in the open field and in forest stands—results of investigations in the state of Hesse. In *Atmospheric Pollutants in Forest Areas* (Ed. H.W. Georgii), pp. 47–67. Reidel, Dordrecht, vii + 287 pp., illus.

Breeman, N. van, & Mulder, J. (1986). Atmospheric acid deposition: effects on the chemistry of forest soils. In T. Schneider, ed. *Acidification and its Policy Implications*, Elsevier, Amsterdam.

Brennan, E., & Leone, I.A. (1969). Suppression of ozone toxicity symptoms in virus-infected tobacco. *Phytopathology*, **59**, 263–264.

Brennan, E.G., Leone, I.A., & Daines, R.H. (1950). Fluoride toxicity in tomato as modified by alterations in the nitrogen, calcium and phosphorus nutrition of the plant. *Plant Physiol.* **25**, 736–747.

Brewer, R.F. (1960). Some effects of hydrogen fluoride gas on seven citrus varieties. *Proc. Am. Soc. Hort. Sci.*, **75**, 236–243.

Brewer, R.F., Guillemet, F.B., & Creveling, R.K. (1960). Influence of N–P–K fertilization on incidence and severity of oxidant injury to mangels and spinach. *Soil Sci.* **92**, 298–301.

Brown, S., & Lugo, A.E. (1984). Biomass of tropical forests: A new estimate based on forest volumes. *Science*, **223**, 1288–1293.

Butin, H., & Wagner, C.H. (1985). Mykologschen untersuchungen zür 'nadelrote' der fichte. *Forstwis. Centrolblatt.*, **104**, 178–186.

California Air Resources Board (1987). Effect of ozone on vegetation and possible alteration of ambient air quality standards. Staff Report, Air Resources Board, Sacramento.

Cameron, J.W., & Taylor, O.C. (1973). Injury to sweet corn inbreds and hybrids by air pollutants in the field and by ozone treatments in the greenhouse. *J. Environ. Qual.* **2**, 387–389.

Carlson, R.W. (1979). Reduction in photosynthetic role of *Acer*, *Quercus* and *Fraxinus* species caused by sulphur dioxide and ozone. *Environ. Pollut.* **18**, 159–170.

Castillo, F.J., & Greppin, H. (1986). Balance between anionic and cationic extracellular peroxidase activities in *Sedum album* leaves after ozone exposure. Physiol. Plantarum, **68**, 201–208.

Cepel, N. (1988) Pollution problems in Turkish forests. Int. Symp. on Plants and Pollutants in Developed and Developing Countries. (Abst.) 22–28 August 1988, Izmir, Turkey.

Chang, C.W. & Thompson, C.R. (1966). Site of fluoride accumulation in navel orange leaves. *Plant Physiol.*, **41**, 211–213.

Chiba, O., & Tanaka, K. (1968). The effect of sulfur dioxide on the development of pine needle blight caused by *Rhizosphaera kalkoffii* Bubac (L.). *J. Jap. Forest Sci.*, **50**, 135–139.

CIAP (1974). *Report of Findings: The Effects of Stratospheric Pollution by Aircraft.* DOT-TSC-75-50, Climate Impact Assessment Program, US Dept. Transportation, Washington, DC.

Cicerone, R.J. (1987). Changes in stratospheric ozone. *Science*, **237**, 35–42.

Clark, B., Henninger, M., & Brennan, E. (1983). An assessment of potato losses caused by oxidant air pollution in New Jersey. *Phytopathology*, **73**, 104–105.

Clark, W.C. (ed.) (1982). *Carbon Dioxide Review 1982*. Clarendon (Oxford University Press), New York.

Colbourne, P., Alloway, B.J., & Thornton, I. (1975). Arsenic and heavy metals in soils associated with regional geochemical anomalies in Southwest England. *Sci. Total Environ.*, **4**, 359–363.

Committee on Trail Smelter Smoke (1939). *Effect of Sulphur Dioxide on Vegetation.* Nat. Res. Council of Canada, Ottawa. 447 pp. illus.

Compton, O.C., & Remmert, L.F. (1960). Effect of airborne fluoride on injury and fluoride content of gladiolus leaves. *Proc. Am. Soc. Hort. Sci.*, **75**, 663–675.

Conkey, L. (1987). Red spruce tree-ring density and growth decline. In G.C. Jacoby & J.W. Hornbeck (eds), *Proc. Internat. Symp. on Ecological Aspects of Tree-ring Analysis*, pp. 382–391, Nat. Tech. Info. Serv., US Dept. Commerce, Springfield, VA.

Cooley, D.R., & Manning, W.J. (1987). The impact of ozone on assimilate partitioning. *Environ. Pollut.* **47**, 95–113.

Costonis, A.C., & Sinclair, W.A. (1969). Ozone injury to *Pinus strobus*. *J. Air Poll. Cont. Assoc.*, **19**, 867–872.

Costonis, A.C., & Sinclair, W.A. (1972). Susceptibility of healthy and ozone injured needles of *Pinus strobus* to invasion by *Lophodermium pinastri* and *Aureobasidium pullularia*. *Eur. J. Forest Pathol.*, **2**, 65–73.

Coulson, C.L., & Heath, R.L. (1975). The interaction of peroxyacetyl nitrate (PAN) with the electron flow of isolated chloroplasts. *Atmos. Environ.*, **9**, 231–238.

Cowling, D.W., & Kozial, M.J. (1982). Mineral nutrition and plant response to air pollutants. In M.H. Unsworth & D.P. Ormrod (eds), *Effects of Gaseous Air Pollution in Agriculture and Horticulture*, pp. 349–375, Butterworths, London.

Craker, L.E. (1971). Ethylene production from ozone-injured plants. *Environ. Pollut.*, **1**, 299–344.

Craker, L.E., & Starbuck, J.S. (1973). Leaf age and air pollutant susceptibility: Uptake of ozone and sulfur dioxide. *Environ. Res.*, **6**, 91–94.

Creer, R.N. (1968). Social psychological factors involved in the perception of air pollution as an environmental health problem. MS Dissertation, University of Utah.

Crittenden, P.D., & Read, D.J. (1979). The effect of air pollution (SO_2) on plant growth. III. Growth studies with *Lolium multiflorum* Lam. and *Dactylis glomerata* L. *New Phytol.*, **83**, 645–651.

Cure, W.W., Sander, J.S., & Heagle, A.S. (1986). Crop yield response predicted with different characterizations of the same ozone treatments. *J. Environ. Qual.*, **15**, 251–254.

Daines, R.H., Leone, I., & Brennan, E. (1952). The effect of fluorine on plants as determined by soil nutrition and fumigation studies. In L.C. McCabe (ed.), *Air Pollution, Proc. US Tech. Conf.*, pp. 97–104, McGraw Hill, New York.

Daines, R.H., Leone, I.A., & Brennan, E. (1960). Air Pollution as it affects agriculture in New Jersey. *New Jersey Agric. Exp. Sta. Bull.*, **794**, 14 pp.

Damicone, J.P., Manning, W.J., & Herbert, S.S. (1987). Growth and disease responses of soybeans from early maturity groups to ozone and *Fusarium oxysporum*. *Environ. Pollut.* **48**, 117–130.

Dana, S.T. (1908). Extent and importance of the white pine blight. USDA Forest Service.

Darley, E.F. (1963, 1964). Festellungen von ozon- und PAN-Wirkung an Pflanzen in der Bundesrepubik Deutschland, Essen (unpublished).

Darley, E.F., & Middleton, J.T. (1966). Problems in air pollution in plant pathology. *Ann. Rev. Phytopath.*, **4**, 103–118.

Darley, E.F., Nichols, C.W., & Middleton, J.T. (1966). Identification of air pollution damage to agricultural crops. *Bull. Calif. Dept. Agric.*, **55**, 11–19.

Darley, E.F., Taylor, O.C., & Middleton, J.T. (1968). Photochemical and exhaust fume damage to plants. 1st International Congress on Plant Pathology, London. 225 pp.

Davis, D.D. (1970). The influence of ozone on conifers. PhD thesis, Pennsylvania State University, University Park, PA.

Davis, D.D., & Wilhour, R.G. (1976). Susceptibility of woody plants to sulfur dioxide and photochemical oxidants. *EPA Ecological Research Series*, EPA-600/3-76-102.

Davis, D.D. & Wood, F.A. (1972). The relative susceptibility of eighteen coniferous species to ozone. *Phytopathology*, **62**, 14–19.

Davis, D.D., & Wood, F.A. (1973). The influences of environmental factors on the sensitivity of Virginia pine to ozone. *Phytopathology*, **63**, 371–376.

Dawson, J.L., & Nash, T.H., III (1980). Effects of air pollution from copper smelters on a desert grassland community. *Environ. Expt. Bot.*, **20**, 61–72.

Dean, C.E., & Davis, D.R. (1972). Ozone air pollution and weather fleck of tobacco. Univ. Florida, Gainsville, Agric. Exp. Sta. Cincinnati, pp. 5–218.

Dean, G., & Treshow, M. (1965). Effects of fluoride on the virulence of tobacco mosaic virus *in vitro*. *Utah Acad. Sci. Arts Ltrs. Proc.* **42**, 236–239.

DeCormis, L., & Luttringer, M. (1977). Influence of sulfur dioxide–nitrogen dioxide synergism. *Pollut. Atmos.*, **75**, 245–247.

Denison, W.C. (1973). Life in tall trees. *Scientific American*, **228**, 74–80.

DeOng, E.R. (1946). Injury to apricot leaves from fluoride deposit. *Phytopathology*, **36**, 469–471.

Detwiler, R.P., & Hall, C.A.S. (1988). Tropical forests and the global carbon cycle. *Science*, **239**, 42–50.

Deumling, D. (1986). Barking up the wrong tree—our mistakes in explaining and combating forest damage. In *Symposium on Acid Deposition*, Muskaku, Ontario, Sept. 1985, pp. 191–202.

Deumling, D. (1987). Airborne chemicals. *Environ. Sci. Technol.*, **21**, 612–613.

Dighton, J., & Skeffington, R.A. (1987). Effects of artificial acid precipitation on the mycorrhizas of Scots pine seedlings. *New Phytol.* **107**, 191–202.

Dochinger, L.S., & Seliskar, C.E. (1970). Air pollution and the chlorotic dwarf disease of eastern white pine. *For. Sci.* **16**, 46–55.

Dochinger, L.S., & Townsend, A.M. (1979). Effects of roadside deicer salts and ozone on red maple progenies. *Environ. Pollut.*, **19**, 229–237.

Dochinger, L.S., Bender, F.W., Box, F.O., & Heck, W.W. (1970). Chlorotic dwarf of eastern white pine caused by ozone and sulphur dioxide interaction. *Nature*, **255**, 476.

Dohler, G. (1984). Effect of uv-b radiation on biomass production, pigmentation, and protein content of marine diatoms. *Z. Naturforsch.*, **39c**, 634–638.

Duggar, W.M., Jr, & Palmer, R.L. (1969). Carbohydrate metabolisms in leaves of rough lemons as influenced by ozone. *Proc. 1st. Int. Citrus Symp.*, **2**, 711–715.

Duggar, W.M., Jr., & Ting L.P. (1968). The effects of peroxyacetyl nitrate on plants. Photoreductive reactions and susceptibility of bean plants to PAN. *Phytopathology*, **56**, 1102–1107.

Duggar, W.M., Jr., Koukel, J., Reed, W.D., & Palmer, R.L. (1963). Effect of peroxyacetyl nitrate on $^{14}CO_2$ fixation by spinach chloroplasts on pinto bean plants. *Plant Physiol.*, **38**, 468–472.

Duggar, W.M., Mudd, J.B., & Koukol, J. (1965). Effect of PAN on certain photosynthetic reactions. *Arch. Environ. Health*, **10**, 195–200.

Dunning, J.A., & Heck, W.W. (1977). Response of bean and tobacco to ozone: Effect of light intensity, temperatures and relative humidity. *J. Air Pollut. Contr. Assoc.*, **27**, 882–886.

Dunning, J.A., Heck, W.W., & Tingey D.T. (1974). Foliar sensitivity of pinto bean and soybean to ozone as affected by temperature, potassium nutrition, and ozone dose. *Water Air Soil Pollut.*, **3**, 305–313.

Eckstein, R.W. von, & Bauch, J. (1983). Dendroklimatologische Unterschungen zum Tanneusterken. *Eur. J. For. Path.*, **13**, 279–288.

Edelman, S. (1968). Air pollution control legislation. In A.C. Stern (ed.), *Air Pollution* V, III, pp. 553–599, Academic Press, New York.

Elkiey, T., & Ormrod, D.P. (1979). Ozone and/or sulfur dioxide effect on tissue permeability in petunia leaves. *Atmos. Environ.*, **13**, 1165–60.

Elkiey, T., & Ormrod, D.P. (1980). Response of turfgrass cultivars to ozone, sulfur dioxide, nitrogen dioxide and their mixtures. *J. Am. Soc. Hort. Sci.*, **105**, 664–668.

Ellenberg, H. (1984). Floristic changes due to eutrophication. In W.A.H. Asman & S.M.A. Riederon (eds), *Ammonia and Acidification*, pp. 301–308, Symp. European Assoc. for Sci. of Air Pollution, Bilthoven, Netherlands, 13–15 April.

Ellenberg, H. (1987). Fulle-Schwund-Schutz: Was will der naturschutz eigentlich? Über Grenzen des Naturschutzes unter den Bedingungen modernen Landnutzung Betriebswirtschaftliche Mitteilungen der Landwirschaftskammer Schleswig Holstein nr. **385**, April, pp. 3–15.

Elliott, C.L., Eberhardt, J.C., & Brennan, E.C. (1987). The effect of ambient ozone pollution and acidic rain on the growth and chlorophyll content of green and white ash. *Environ. Pollut.*, **44**, 66–70.

Engle, R.L., & Gabelman, W.H. (1966). Inheritance and mechanisms for resistance to ozone damage in onion (*Allium cepa* L.). *J. Am. Soc. Hort. Sci.*, **89**, 423–430.

Engle, R.L., Gabelman, W.H., & Romanowski, R.R., Jr. (1965). Tipburn, an ozone incited response in onion (*Allium cepa L.*). *J. Am. Soc. Hort. Sci.*, **86**, 468–474.

Facteau, T.J., Wang, S.Y., & Rowe, K.E. (1973). The effect of hydrogen fluoride on pollen germination and pollen tube growth in *Prunus avium* L. cv 'Royal Ann'. *J. Am. Soc. Hort. Sci.*, **98**, 234–236.

Facteau, T.J., & Rowe, K.E. (1977). Effect of hydrogen fluoride and hydrogen chloride on pollen-tube growth and sodium fluoride on pollen germination in 'Tilton' apricot. *J. Amer. Soc. Hort. Sci.*, **102**, 95–96.

Farnworth, E.G., Tidrick, T.H., Jordan, C.F., & Smathers, W.M. Jr. (1981). The value of natural ecosystems; an economic and ecological framework. *Environ. Conserv.*, **8**, 275–282.

Fassett, D.W. (1975). Cadmium: Biological effects and occurrence in the environment. *Ann. Rev. Pharmacol.*, **15**, 425–435.

FBW (1986). Research Advisory Council, Second Report.

Federal Ministry of Food, Agriculture and Forestry (1984). 1984 Forest Damage Survey. Bonn, p. 22.

Ferenbaugh, R.W. (1976). Effects of simulated acid rain on *Phasealus vulgaris* L. (*Fabaceae*). *Am. J. Bot.*, **63**, 283–288.

Ferm, A., Hytonen, J., Kolari, K.K., & Veijalainen, H. (1987). Growth disturbances of forest trees close to fur animal farms. In W.A.H. Asman & H.S.M.A. Diederen (eds), *Symp. of European Assoc. for Science of Air Pollution*, p. 309, Nat. Inst. Public Health & Env. Hygiene, Bilthoven, Netherlands, 13–15 April.

Ferry, B.W., Baddeley, M.S., & Hawksworth, D.L. (eds) (1973). *Air Pollution and Lichens*, Athlone Press, London.

Fiedler, H.J., & Hohne, H. (1984). Die Bor-Enahrung von Koniferen und ihre Beziehung züm Gehalt an Calcium und Kalium in den Assimilations organen. *Beitrage Forstwirtschaft*, **18**, 73–80.

Filner, P., Rennenberg, H., Sekiya, J., Bressan, R.A., Wilson, L.G., Curedux, L. Le, & Shimei, T. (1984). Biosynthesis and emission of hydrogen sulfide by higher plants. In M. Koziol & F.R. Whatley, *Gaseous Air Pollutants and Plant*

Metabolism, pp. 291–312, Butterworths, London.

Fluhler, H., Polomski, J., & Blaser, P. (1981). Dynamik der Fluorakkumulation in einem immissionung belasteten Boden. *Mittg. Deutsch. Bodenkundl. Gessellsch.*, **30**, 31–92.

Forberg, E., Aarnes, H., & Nisen, S. (1987). Effect of ozone on net photosynthesis in oat (*Avena sativa*) and duckwood (*Lemna gibba*). *Environ. Poll.*, **47**, 285–291.

Foster, K.W., Guerard, V.P., Oshima, R.J., & Timm, H. (1983). Differentiatial ozone susceptibility of centennial russet and white rose potato as demonstrated by fumigations and antioxidant treatments. *Ann. Potato J.*, **60**, 127–139.

Fox, C.A., Kinkaid, W.B., Nash, T.H. III, Young, D.L., & Fritts, H.C. (1986). Tree-ring variation in western larch (*Larix occidentalis*) exposed to sulfur dioxide emissions. *Can. J. For. Res.*, **16**, 283–292.

Fretag, M. (1869). Mitt. d. Konigl. landw. Akad. Popplsdorf, 2. (Cited in Sorauer, 1914).

Friedland, A.J., Gregory, R.A., Karenlampi, L., & Johnson, A.H. (1984). Winter damage to foliage as a factor in red spruce decline. *Can. J. For. Res.*, **14**, 963–965.

Gardner, J.O., & Ormrod, D.P. (1977). Response of Reiger begonia to ozone and sulphur dioxide. *Sci. Hort.*, **5**, 171–181.

Glater, R.B., Solberg, R.A., & Scott, F.M. (1962). A developmental study of the leaves of *Nicotiana glutinosa* as related to their smog sensitivity. *Am. J. Bot.*, **49**, 954–970.

Godish, T. (1985). *Air Quality*, Lewis Publ., Inc. Chelsea, MI.

Godzik, S., & Sassen, M.M.A. (1974). Einwirkung von SO_2 auf die feinstruktur der Chloroplasten von *Phaseolus vulgaris*. *Phytopath. Z.*, **79**, 155–159.

Goldman, J.C., & Dennett, M.R. (1983). Carbon dioxide exchange between air and seawater: No evidence for rate catalysis. *Science*, **220**, 199–201.

Goodman, R.N., Kiraly, Z., & Zaitlin, M. (1976). *The Biochemistry and Physiology of Infectious Plant Diseases*, Van Nostrand. Princeton.

Grzywacz, A., & Wazny, J. (1973). The impact of industrial air pollutants on the occurrence of several important pathogenic fungi of forest trees in Poland. *Eur. J. For. Path.*, **3**, 129–141.

Guderian, R. (1977a). Air pollution. In *Ecological Studies* 22, p. 127, Springer-Verlag, Berlin.

Guderian, R. (1977b). Phytotoxicity of acidic gases and its significance in air pollution control. Ecol. Stud. 22. *Air Pollution*, Springer Verlag, Berlin, 127 pp.

Guderian, R., & Stratmann, H. (1968). *Forschungsbeichte des Landes Nordrhein-Westfalen*, No. 1920, 114 pp.

Guderian, R., Tingey, D.T., & Rabe, R. (1985). Effects of photochemical oxidants on plants. In R. Guderian (ed.), *Air Pollution by Photochemical Oxidants*, pp. 130–346, Springer-Verlag, Berlin.

Guderian, R., Klumpp, G., & Klumpp, A. (1988). Effects of SO_2, O_3, NO_+, singly and in combination on forest species. Internat. Symp. on Plants and Pollutants in Developed and Developing Countries, 22–28 August, p. 33.

Gusten, H., Heinrich, G., Cvitas, T., Klasine, L., Ruscie, B., Lalas, D.P., & Petrakis, M. (1988). Photochemical formation and transport of ozone in Athens, Greece. *Atmos. Environ.*, **22**, 1855–1861.

Haagen-Smit, A.J. (1952). Chemistry and Physiology of Los Angeles smog. *Ind. Eng. Chem.*, **344**, 134–134.

Haagen-Smit, A.J., Darley, E.F., Zaitlin, M., Hull, H., & Noble, W. (1952). Investigations on injury to plants from air pollution in the Los Angeles area. *Plant Physiol.*, **27**, 18–34.

238   *References*

Halbwachs, G. (1968). Untersuchungen über den Wasserhaushalt rauchgeschadgter Forstgeholze. In *Referaten der VI. Interantionden Arbeitstagung forstliches Rauchschadensachverst in giger*, pp. 209–281, Polskiej Akademi Nouk Katowice 9–14 September.

Halbwachs, G. (1988). Die Belastung der Wälder—ein multifaktorielles Problem. *Oesterrechische Wasserwirtschaft*, **40**, 114–118.

Hale, M.E., Jr (1961). *Lichen Handbook: A Guide to the Lichens of Eastern North America*, Smithsonian Institution Press, Washington, DC, 178 pp.

Hale, M.E., Jr (1967). *The Biology of Lichens*, Arnold, London, 176 pp.

Hale, M.E., Jr (1969). *How to Know the Lichens*, Brown, Dubuque, IA, 226 pp.

Haliday, E.C. (1961). A historical review of atmospheric pollution. In *Air Pollution*, pp. 9–35, WHO, Columbia University Press, New York.

Hallgren, J.E. (1978). Physiological and biochemical effects of sulphur dioxide on plants. In J.O. Nriagu (ed.), *Sulphur in the Environment*: Part II *Ecological Impact*, pp. 163–209, Wiley, New York.

Hansen, J., Gornitz, V., Lebedeff, S., & Moore, E. (1983). Global mean sea level: Indicator of climatic change? [Letter response to an article by R. Etkins & E. Epstein, *Science*, **215**: 287 (1982)]. *Science*, **219**, 996–998.

Hanson, G.P., & Stewart, W.S. (1970). Photochemical oxidants: Effects on starch hydrolysis in leaves. *Science*, **168**, 1223–1224.

Harkov, R., & Brennan, E. (1979). An ecophysiological analysis of the response of trees to oxidant pollution. *J. Air Pollut. Contr. Assoc.*, **29**, 157–161.

Harper, J.L. (1977). *Population Biology of Plants*, Academic Press, London, 892 pp.

Haut, H. van, & Stratmann, H. (1970). *Farbtafelatlas über Schwefeldioxid-Wirkungen an Pflanzen*, Girardet, Essen, 206 pp.

Hartman, F.E. (1924). The industrial application of ozone. *J. Am. Soc. Heat. Vent. Eng.*, **30**, 711–727.

Hartogensis, F. (1979). Criteria for establishing legislation, regulations and planning guidelines concerning ambient concentrations of air-borne pollutant. UN Economic Commission for Europe, Symposium, 8 June.

Harvey, G.W., & Legge, A.A. (1979). The effect of sulphur dioxide upon the metabolic level of adenosine triphosphate. *Can. J. Bot.*, **57**, 759–764.

Hasselhoff, E., & Lindau, G. (1903). *Handbuch zur Erkennung und Beurteilung von Rauschäden*. Gebrüder Borntraeger, Berlin, 315 pp.

Hauhs, M., & Wright, R.F. (1986). Regional pattern of acid deposition and forest decline along a cross section through Europe. *Air Water Soil Pollution.*, **31**, 463–474.

Havas, P. (1971). Injury to pines in the vicinity of a chemical processing plant in northern Finland. *Acta Forestalia Fennica*, **121**, 1–21.

Hawksworth, D.L. (1973). Lichens as litmus for air pollution: a historical review. *Int. J. Env. Studies*, **1**, 281–296.

Haywood, J.K. (1905). Injury to vegetation by smelter fumes. *US Dept. Agric. Bur. Chem. Bull. No. 89*, 23 pp.

Haywood, J.K. (1910). Injury to vegetation and animal life by smelter wastes. *US Dept. Agric. Bur. Chem. Bull. No. 113*, 63 pp.

Heagle, A.S. (1972). Effect of ozone and sulfur dioxide on injury, growth, and yield of soybean. *Phytopathology*, **62**, 763 (Abstract).

Heagle, A.S. (1973). Interactions between air pollutants and plant parasites. *Ann. Rev. Phytopathol.*, **11**, 365–388.

Heagle, A.S. (1979). Effects of growth media, fertilizer rate and hour and season of exposure on sensitivity of four soybean cultivars to ozone. *Environ. Pollut.*, **8**, 313–322.

Heagle, A.S., Heck, W.W., Lesser, V.M., & Rawlings, J.O. (1987). Effects of daily ozone exposure duration and concentration fluctuation in yield of tobacco. *Phytopathology*, **77**, 856–862.

Heagle, A.S., & Johnston, J.W. (1979). Variable response of soybeans to mixtures of ozone and sulfur dioxide. *J. Air Poll. Contr. Assoc.*, **29**, 729–732.

Heagle, A.S., & Key L. (1973). Effect of *Puccinia graminis* f. sp. *tritici* on ozone injury in wheat. *Phytopathology*, **63**, 609–613.

Heagle, A.S., & Philbeck, R.B. (1978). Exposure techniques. In W.W. Heck, S.V. Krupa, & S.N. Linzon (eds), *Methodology for the Assessment of Air Pollution Effects on Vegetation*, Ch. 6, pp. 1–19, Air Pollution Contr. Assoc., Pittsburgh.

Heagle, A.S., Body, D.E., & Neely, G.E. (1974). Growth and yield responses of soybean to chronic doses of ozone and sulfur dioxide in the field. *Phytopathology*, **64**, 1372–1376.

Heagle, A.S., Philbeck, R.B., & Letchworth, M.B. (1979). Injury and yield responses of spinach cultivars to chronic doses of ozone in open-top field chambers. *J. Environ. Qual.*, **8**, 368–373.

Heagle, A.S., Spencer, S., & Letchworth, M.B. (1980). Yield responses of winter wheat to chronic doses of ozone. *Can. J. Bot.* **57**, 1999–2005.

Heagle, A.S., & Strickland, A. (1972). Reaction of *Erisiphe graminis* f. sp. *hordei* to low levels of ozone. *Phytopathology*, **62**, 1144–1148.

Heath, R.L. (1975). Ozone. In J.B. Mudd & T.T. Kozlowski (eds), *Responses of Plants to Air Pollution*, pp. 23–25. Academic Press, New York.

Heck, W.W., & Pires, E.G. (1962). Effect of ethylene on horticultural and agronomic plants. Texas Agric. Exp. Sta., MP-613, 12 pp.

Heck, W.W., Dunning, J.A., & Hindawi, C.J. (1965). Interactions of environmental factors on the sensitivity of plants to air pollution. *J. Air Poll. Contr. Assoc.*, **15**, 511–515.

Heck, W.W., Krupa, S.V., & Linzon, S.N. (1978). *Methodology for the Assessment of Air Pollution Effects on Vegetation*. Air Pollution Control Association, Pittsburgh, 302 pp.

Heck, W.W., Taylor, O.C., Adams, R., Bingham, G., Miller, J., Preston, E., & Weinstein L. (1982). Assessment of crop loss to ozone. *J. Air Poll. Contr. Assoc.*, **32**, 353–361.

Heck, W.W., Cure, W.W., Rawlings, J.O., Zaragoza, L.J., Heagle, A.S., Haggestad, H.E., Kohut, R.J., Kress, L.W., & Temple, P.J. (1984). Assessing impacts of ozone on agricultural crops. II. Crop yield functions and alternative exposure statistics. *J. Air Poll. Contr. Assoc.*, **34**, 810–817.

Hedgecock, G.G. (1914). Injuries by smelter smoke in southeastern Tennessee. *Wash. Assoc. Sci. J.*, **4**, 70–71.

Heggestad, H.E. (1988). Reduction in soybean seed yields by ozone air pollution. *J. Air Poll. Contr. Assoc.*, **38**, 1040–1041.

Heggestad, H.E., & Bennett, J.H. (1981). Photochemical oxidant potentiate yield losses in snap beans attributable to sulfur dioxide. *Science*, **213**, 1008–1010.

Heggestad, H.E., & Middleton, J.T. (1959). Ozone in high concentrations as a cause of tobacco leaf injury. *Science*, **129**, 208–210.

Heil, G.W., & Diemont, W.M. (1983). Raised nutrient levels change heathland into grassland. *Vegetatio*, **53**, 113–120.

Heil, G.W., Weger, M.J.A., de Mol, W., Van Dam, D., & Heijne, B. (1988). Capture of atmospheric ammonium by grassland canopies. *Science*, **239**, 764–765.

Heitschmidt, R.K., Lauenroth, W.K., & Dodd, J.C. (1978). Effects of controlled levels of sulfur dioxide on western wheatgrass in a southeastern Montana grassland. *J. Appl. Ecol.*, **14**, 859–868.

Heliotis, F.D., Karandinos, M.G., & Whiton, J.C. (1988). Air pollution and the decline of fir forests in Parnas National Park, near Athens, Greece. *Environ. Pollut.*, **54**, 29–40.

Hepting, G.H., & Berry, C.R. (1961). Differentiating needle blights of white pine in the interpretation of fume damage. *Intern. J. Air Water Pollut.*, **4**, 101–105.

Hill, A.C. (1969). Air quality standards for fluoride vegetation effects. *J. Air Poll. Contr. Assoc.*, **19**, 331–336.

Hill, A.C. (1971). Vegetation: A sink for atmospheric pollutants. *J. Air Poll. Contr. Assoc.*, **21**, 341–346.

Hill, A.C., & Littlefield, N. (1969). Ozone: Effect on apparent photosynthesis, rate of transpiration, and stomatal closure in plants. *Environ. Sci. Technol.*, **3**, 52–56.

Hill, A.C., Pack, M.R., Treshow, M., Downs, R.J., & Transtrum, L.G. (1961). Plant injury induced by ozone. *Phytopathology*, **51**, 356–363.

Hill, A.C., Heggestad, H.E., & Linzon, S.N. (1970). Ozone. In J.S. Jacobson & A.C. Hill (eds). *Recognition of Air Pollution Injury to Vegetation: A Pictorial Atlas*, pp. B1–B22, Air Pollution Control Association, Pittsburgh.

Hill, A.C., Hill, S., Lamb, C., & Barrett, T.W. (1974). Sensitivity of native desert vegetation to SO_2 and to SO_2 and NO_2 combined. *J. Air Poll, Contr. Assoc.*, **24**, 153–157.

Hoffman, G.A., Maas, E.V., & Rawlins, S.L. (1975). Salinity–ozone interactive effects on alfalfa yield and water relations. *J. Environ. Qual.*, **4**, 326–331.

Hofmann, G. (1955). Zur Darstellung der spektralen Verteilung der Strahlungsenergie. *Arch. Met. (B)*, **6**, 274–279.

Hofstra, G., Littlejohns, D.A., & Wukasch, R.T. (1978). The efficacy of the antioxidant ethylene-diurea (EDU) compared to carboxin and benomyl in reducing yield losses from ozone in navy bean. *Plant Dis. Rep.*, **62**, 350–352.

Homan, C. (1937). Effects of ionized air and ozone on plants. *Plant Physiol.*, **12**, 937.

Howard, A.L. (1948). *A Manual of the Timbers of the World: Their Characteristics and Uses* 3rd edn, Macmillan, London, 751 pp.

Howell, R.K. (1974). Phenols, ozone and their involvement in pigmentation and physiology of plant injury. In M. Duggar (ed.), *Air Pollution Effects on Plant Growth*, pp. 94–105, ACS Symp. Ser. 3, Am. Chem. Soc., Washington, DC.

Hudson, R.D., & Reed, E.I. (eds) (1979). The stratosphere: present and future. In *Halocarbons: Effects of Stratospheric Ozone*, NASA Ref. Publ. 1049, National Academy of Science, Washington, DC.

Hull, H.M., & Went, F.W. (1952). Life processes of plants as affected by air pollution. In *Proc. 2nd Nat. Air Pollut. Symp.*, pp. 122–128, Pasadena, CA.

Hursh, C.R. (1948). Local climate in the Copper Basin of Tennessee as modified by the removal of vegetation. *USDA Circ.*, **774**, 38 pp.

Huttunen, S. (1978). The effects of air pollution on provenances of Scots Pine and Norway Spruce in northern Finland. *Silva Fenn.*, **12**, 1–16.

Huttunen, S. (1984). Interactions of disease and other stress factors with atmospheric pollution. In M. Treshow (ed.), *Air Pollution and Plant Life*, pp. 321–356, Wiley, Chichester.

Huttunen, S., & Laine, K. (1981). The structured pine needle surface (*Pinus sylvestris* L.) and the deposition of airborne pollutants. *Archiium Ochromy Srodowiska*, **24**, 29–38.

Irving, P.M. (1987). Gaseous pollutant and acidic rain impacts on crops in the United States: A comparison. *Environ. Tech. Ltrs.*, **8**, 451–458.

Irving, P., Miller, J.E., & Yerikos, P.B. (1982). The effect of NO$_2$ and SO$_2$ alone and in combination on the productivity of field-grown soybeans. In T. Schneider & L. Grand (eds), *Air Pollution and Nitrogen Oxides*, pp. 521–531. Elsevier Scientific, Amsterdam.

Jacobson, J.S., & Colavito, L.J. (1976). The combined effects of sulfur dioxide and ozone on bean and tobacco plants. *Environ. Exp. Bot.*, **16**, 277–285.

Jacobson, J.S. (1977). The effects of photochemical oxidants on vegetation. In *Ozon und Begleit Substanzen im Photochemischen Smog*, pp. 163–173, VD1 Colloquium, 1976, Dusseldorf (VDI-Berichte No. 270).

Jacobson, J.S., & Hill, A.C. (1970). *Recognition of Air Pollution Injury to Vegetation*, Air Pollution Control Association, Pittsburgh.

Jaffe, L.S. (1966). Effects of photochemical air pollution on vegetation. *59th Ann. Mtg. Air Poll. Control Assoc. Paper*, 1966, **43**.

Jancarik, V. (1961). Vyskyt drevokazngch hub v kourem poskozovani oblasti krusngch hor. *Lesnictui*, **7**, 667–692.

Jensen, K.F., & Roberts, B.R. (1986). Changes in yellow poplar stomatal resistance with SO$_2$ and O$_3$ fumigation. *Environ. Pollut. Ser. A.* **41**, 235–245.

Johnson, A.H., Siccama, T.G., Wang, D., Turner, R.S., & Barringer, T.H. (1981). Recent changes on tree growth in the New Jersey Pine Barrens: A possible effect of acid rain. *J. Environ. Qual.*, **10**, 427–430.

Johnson, A.H., Freidland, A.J., & Siccama, T.C. (1983). Recent changes in the growth of forest trees in the northeastern United States. In D.D. Davis, A.A. Miller, & L. Dochinger (eds), *Air Pollution and the Productivity of the Forest*, pp. 121–142, Isaac Walton League, Washington, DC.

Johnson, A.H., Freidland, A.J., & Dushoff, J.G. (1986). Recent and historic red spruce mortality: Evidence of climatic influence. *Water Air Soil Pollution*, **30**, 319–330.

Jones, H.C., Weatherford, F.P., Noggle, J.C., Lee, N.T., & Cunningham, J.R. (1979). Power-plant siting: Assessing SO$_2$ effects on agriculture. Air Pollution Control Association, Pittsburgh, *72nd Annual Meeting Cincinnati, USA*, preprint No. 79–13.5.

Juhren, M., Noble, W., & Went, F.W. (1957). The standardization of *Poa annua* as an indicator of smog concentrations. I. Effects of temperature, photoperiod, and light intensity during growth of test plants. *Plant Physiol.*, **32**, 576–586.

Karhu, M., & Huttenen, S. (1986). Erosion effects of air pollution in needle surfaces. *Air Water Soil Pollution*, **31**, 417–423.

Karnosky, D.F. (1976). Threshold levels for foliar injury to *Populus tremuloides* by sulfur dioxide and ozone. *Can. J. For. Res.*, **6**, 166–169.

Keen, T., & Taylor, O.C. (1975). Ozone injury in soybeans. Isoflavonoid accumulation is related to necrosis. *Plant Physiol.*, **55**, 731–733.

Keller, T. (1977). Der Enfluss von Fluorimmissionen auf die Nettoassimilation von Waldbaumen. *Mitt. Eidg. Anstalt Forst Versuchsw.*, **53**, 161–198.

Keller, T. (1980). The simultaneous effect of soil-borne NaF and air pollutant SO$_2$ on CO$_2$-uptake and pollutant accumulation. *Oecologia*, **44**, 283–285.

Kelley, N.A., Wolff, G.T., & Ferman, M.Z.A. (1982). Background pollutant measurements in air masses affecting the eastern half of the United States. I. Air masses arriving from the northwest. *Atmos. Environ.*, **16**, 1077–1088.

Kelley, N.A., Wolff, G.T., & Ferman, M.A. (1984). Sources and sinks of ozone in rural areas. *Atmos. Environ.*, **18**, 1251–1266.

Kender, W.J., & Shaulis, N.J. (1976). Vineyard management practices influencing oxidant injury on concord grapevines. *J. Am. Soc. Hort. Sci.*, **101**, 129–132.

Kerr, R.A. (1983). Trace gases could double climate warming (A research news report). *Science*, **220**, 1364–1365.

King, D. (1987). Influence of moisture stress on crop sensitivity to ozone. NCLAN Conference Paper, Raleigh, NC, October 28.

Kinkaid, W.B. (1987). Dendrochronological analysis of ambient sulfur dioxide effects in western larch. In G.C. Jacoby Jr & J.W. Hornbeck (eds), *Proc. Internat. Symp. on Ecological Aspects of Tree Ring Analysis*, pp. 410–416, Nat. Tech. Info. Serv., Springfield, VA.

Kirschner, M. (1987). Klima, Witterung und Waldschäden. *Inhält*, **Sept.**, 39–44.

Klarer, C.I. (1982). Effects of sulfur dioxide and nitrogen dioxide, singly and in combination over time on ribulose bisphosphate carboxylase activity and vegetative growth of soybean. MS thesis, North Carolina State University Raleigh, NC, 76 pp.

Klein, R.M., & Perkins, T.D. (1987). Cascades of causes and effects of forest decline, *Ambio.*, **16**, 86–93.

Knabe, W., Brandt, C.S., Haut, H. Van, & Brandt, C.J. (1973). Nachuseis photochemischer Luftverunreinigunger durch biologische Indikatoren in den Bundesrepublik Deutschland. *Proc. 3rd Int. Clean Air Congr.*, VDI-Verlag, Dusseldorf, A110–A114.

Knudson, L.L., Tibbets, T.W., & Edwards, G.E. (1977). Measurement of ozone injury by determination of leaf chlorophyll concentration. *Plant Physiol.*, **60**, 606–608.

Kohut, R.J. (1972). Response of hybrid poplar to simultaneous exposures to ozone and PAN. MS thesis. Pennsylvania State University. CAES Publ. 288–72, Cent. Env. Studies, Pennsylvania State University State College.

Kohut, R.J., & Davis, D.D. (1978). Response of pinto bean to simultaneous exposure to ozone and peroxyacetyl nitrate. *Phytopathology*, **68**, 567–569.

Kok, G.L. (1980). Measurements of hydrogen peroxide in rainwater. *Atmos. Environ.*, **4**, 653–656.

Kolari, K.K. (1983). *Growth Disturbances of Forest Trees*. Comm. Inst. For. Fenn., No. **116**, Helsinki, 208 pp.

Kontic, R., Niederer, M. Nippel, C., & Winkler-Seifert, A. (1987). Annual ring analysis on conifers for the description and assessment of forest damage (The Valais, Switzerland). In G.C. Jacoby & J.W. Hornbeck (eds), *Proc. Internat. Symp. on Ecological Aspects of Tree-ring Analysis*, pp. 417–423, US Dept. Commerce.

Kopp, R.J., Vaughn, W.T., & Hazilla, M. (1984). *Agricultural Benefits Analysis: Alternate Ozone and Photochemical Oxidant Standards*. Final report to Economic Analysis Branch, US EPA, Research Triangle Park, NC, June.

Kopp, R.J., Vaughn, W.J., Hazilla, M., & Carson, R. (1985). Implications of environmental policy for U.S. agriculture. The case of ambient ozone standards. *J. Environ. Manag.*, **20**, 321–331.

Kormelink, J.R. (1967). Effects of ozone on fungi. MS thesis, University of Utah, Salt Lake City, 28 pp.

Koziol, M.J. (1984). Interactions of gaseous pollutants with carbohydrate metabolism in M.J. Koziol & F.R. Whatley (eds), *Gaseous Air Pollutants and Plant Metabolism*, pp. 251–273, Butterworths, London.

Koziol, M.J., & Jordan, C.E. (1978). Changes in carbohydrate levels in red kidney bean (*Phaseolus vulgaris* L.) exposed to sulphur dioxide. *J. Exp. Bot.*, **29**, 1037–1043.

Krause, C.R., & Weidensaul, T.C. (1978). Effects of ozone on the sporulation, germination, and pathogenicity of *Botrytis cinerea*. *Phytopathology*, **68**, 195–198.

Krause, G.H.M., & Kaiser, H. (1977). Plant response to heavy metals and sulphur dioxide. *Environ. Pollut.*, **12**, 63–71.

Krause, G.H.M. (1986). Zur Wirkung von Ozon und 'Saurem Nebel' auf die Auswaschung von Nahrstoffen aus Fichten. In H. Stratmann, *Au der Tätigkeit der LIS*, pp. 43–46, Landesanstat für Immissionsschutz des Landes Nordrein-Westfalen, Essen.

Kress, L.W. (1972). *Response of Hybrid Poplar to Sequential Exposures of Ozone and PAN*. MS thesis. Pennsylvania State University, University Park, PA, 39 pp.

Kress, L.W. (1980). Effect of O_3 and O_3 + NO_2 on growth of tree seedlings. p 239. In *Proc. Symp. of Effects of Air Pollutants on Mediterranean and Terrestrial Forest Ecosystems*, Rept. PSW-43, Pacific Southwest Forest and Range Exp. Sta., Berkeley, CA, 256 pp.

Kress, L.W., & Skelly, J.M. (1982). Response of several eastern forest tree species to chronic doses of ozone and nitrogen dioxide. *Plant Dis.*, **66**, 1149–1152.

Kress, L.W., Skelly, J.M., & Hinkelman, K.H. (1982a). Growth impact of O_3, NO_2 and/or SO_2 on *Pinus taeda*. *Environ. Monit. Assoc.*, **1**, 229–239.

Kress, L.W., Skelly, J.M., & Hinkelman, K.H. (1982b). Growth impact of O_3, NO_2 and/or SO_2 on *Platanus occidenalis*. *Agric. Environ.*, **7**, 265–274.

Kuntasal, G., & Chang, T.Y. (1987). Trends and relationships of O_3, NO_x, and HC in the South Coastal Air Basin of California. *J. Air Poll. Contr. Assoc.*, **37**, 1158–1163.

Lacasse, N.L., & Treshow, M. (1976). *Diagnosing Vegetation Injury Caused by Air Pollution*, Environmental Protection Agency. Washington, DC, 139 pp.

Lai Dinh, D., Buchloh, G., & Oelschlager, W. (1973). Auswirkung von Fluorverbindungen auf die Pollenkeuning und den Frucktansatz von Obstgewächsen. *Der Erwerbsobst Bau.*, **15**, 154–157.

LaMarche, V.C., Jr, Graybill, D.A., Fritts, H.C., & Rose, M.R. (1984). Increasing atmospheric carbon dioxide: Tree ring evidence for growth enhancement in natural vegetation. *Science*, **225**, 1019–1021.

Lamoreaux, R.J., & Chaney, W.R. (1978). Photosynthesis and transpiration of excised silver maple leaves exposed to cadmium and sulphur dioxide. *Environ. Pollut.*, **17**, 259–268.

Lauenroth, W.K., & Heasley, I.E. (1980). Impact of atmospheric sulfur deposition on grassland ecosystems. In D.S. Shriner, C.R. Richinard, & S.E. Lindberg (eds), *Atmospheric Sulfur Deposition*, pp. 417–430, Ann Arbor Sci. Publ., Ann Arbor, MI.

LeBlanc, F., & Rao, D.N. (1975). Effects of air pollutants on lichens and bryophytes. In J.B. Mudd & T.T. Kozlowski (eds), *Responses of Plants to Air Pollution*, pp. 237–272, Academic Press, New York.

LeBlanc, D.C., Raynal, D.J., White, E.H., & Ketchledge, E.H. (1987a). Characterization of historical growth patterns in declining red spruce trees. In G.C. Jacoby & J.W. Hornbeck (eds), *Proc. Internat. Symp. on Ecological Aspects of Tree-Ring Analysis*, pp. 360–371, Nat. Techn. Info. Serv., US Dept. Commerce, Springfield, VA.

LeBlanc, D.C., Raynal, D.J., & White, E.H. (1987b). Acidic deposition and tree growth. I. The use of stem analysis to study historical growth patterns. *J. Environ. Qual.*, **16**, 325–333.

Ledbetter, M.C., Zimmerman, P.W., & Hitchcock, A.E. (1959). The histopathological effects of ozone on plant foliage. *Contrib. Boyce Thompson Inst.*, **20**, 275–282.

Ledbetter, M.C., Mavrodineanu, R., & Weiss, A.J. (1960). Distribution studies of radioactive fluorine-18 and stable fluoride-19 in tomato plants. *Contrib. Boyce Thompson Inst.*, **20**, 331–348.

Lee, E.H., Tingey, D.T., & Hogsett, W.E. (1987). Evaluation of exposure indices in exposure–response modeling. Int. Conf. Assessment of Crop Loss from Air Pollutants, Raleigh, NC, October 25–29 (to be published in *Env. Poll.*, 1988).

Lee, T.T. (1965). Sugar content and stomatal width as related to ozone injury in tobacco leaves. *Can. J. Bot.*, **43**, 677–685.

Lee, T.T. (1967). Inhibition of oxidative phosphorylation and respiration by ozone in tobacco mitochondria. *Plant Physiol.*, **42**, 691–696.

Lefohn, A.S., & Jones, C.K. (1986). The characterization of ozone and sulfur dioxide air quality data for assessing possible vegetation effects. *J. Air Poll. Contr. Assoc.*, **36**, 1123–1129.

Lefohn, A.S., & Mohnen, V.A. (1986). The characterization of ozone, sulfur dioxide, and nitrogen dioxide for selected monitoring sites in the Federal Republic of Germany. *J. Air Poll. Contr. Assoc.*, **36**, 1329–1337.

Lefohn, A.S., & Ormrod, D.P. (eds) (1984). A *Review and Assessment of the Effects of Pollutant Mixtures on Vegetation*, EPA-60013-84-037, US EPA, Corvallis, OR. 104 pp.

Lefohn, A.S., Lawrence, J.A., & Kohut, A.J. (1988). A comparison of indices that describe the relationship between exposure to ozone and reduction in the yield of agricultural crops. *Atmos. Environ.* **22**, 1229–1240.

Legassicke, B.C., & Ormrod, D.P. (1981). Suppression of ozone-injury on tomatoes by ethylene diurea in controlled environment and in the field. *Hort. Sci.*, **16**, 183–184.

Lemon, E.R. (ed.) (1983). *CO₂ and Plants: The Response of Plants to Rising Levels of Atmospheric Carbon Dioxide*. AAAS Selected Symposia Series No. 84, Proceedings of a conference at Athens, Georgia, May 1982, published for the American Association for the Advancement of Science by Westview, Boulder, CO, 280 pp.

Leonard, C.D., & Graves, H.B. (1966). Effect of airborne fluoride on Valencia orange yields. *Proc. Fla. State Hort. Soc.*, **79**, 79–86.

Leonard, H.J. (1985). Confronting industrial pollution in rapidly industrializing countries. Myths, pitfalls and opportunities. *Ecol. Law Quarterly*, **12**, 779–816.

Leone, I.A. (1976). Response of potassium-deficient tomato plants in atmospheric ozone. *Phytopathology*, **66**, 734–736.

Leone, I.A., & Brennan, E. (1970). Ozone toxicity in tomato as modified by phosphorus nutrition. *Phytopathology*, **60**, 1521–1524.

Leone, I.A., & Brennan, E. (1972). Modification of sulfur dioxide injury to tobacco and tomato by varying nitrogen and sulfur nutrition. *J. Air Pollut. Contr. Assoc.*, **22**, 544–547.

Leone, I.A., Brennan, E., & Daines, R.H. (1966). Effect of nitrogen nutrition on the response of tobacco to ozone in the atmosphere. *J. Air Poll. Contr. Assoc.*, **16**, 191–196.

Lepp, N.W. (1980). *Effects of Heavy Metals Pollution on Plants*, Applied Science, New York, 352 pp.

Leung, S.K., Reed, W., & Geng, S. (1982). Estimation of ozone damage to selected crops grown in southern California. *J. Air Poll. Contr. Assoc.*, **32**, 160–164.

Levitt, J. (1980). *Responses of Plants to Environmental Stresses*, 2nd ed, Academic Press, London, 697 pp.

Lewis, E., & Brennan, E. (1978). Ozone and sulfur dioxide mixtures cause a PAN-type injury to petunia. *Phytopathology*, **68**, 1011–1014.

Likens, G.E., & Borman, F.H. (1974). Acid rain: A serious regional environmental problem. *Science*, **184**, 1176–1179.

Lindberg, S.E., Lovett, G.M., Turner, R.R., & Hoffman, W.H. (1985). Atmospheric chemistry deposition and canopy interactions. In D.W. Johnson (ed.), *Walker Branch Watershed Synthesis*, New York, Springer-Verlag.

Linzon, S.N. (1960). The development of foliar symptoms and the possible cause of white pine needle blight. *Can. J. Bot.*, **38**, 153–161.

Linzon, S.N. (1966). Damage to eastern white pine by sulfur dioxide, semimature tissue needle blight and ozone. *J.Air Poll. Contr. Assoc.*, **16**, 140–144.

Linzon, S.N. (1967). Ozone damage and semi-mature tissue needle blight of eastern white pine. *Can. J. Bot.*, **45**, 2047–2061.

MacDowall, F.D.H., & Cole, A.F.W. (1971). Threshold and synergistic damage to tobacco by ozone and sulfur dioxide. *Atmos. Environ.*, **5**, 553–559.

MacDowall, F.D.H., Vickery, L.S., Runecles, V.C., & Patrick, Z.A. (1963). Ozone damage to tobacco in Canada. *Can. Plant Dis. Survey*, **43**, 131–151.

MacDowall, F.D.H. (1965). Predisposition of tobacco to ozone damage. *Can. J. Plant Sci.*, **45**, 1–12.

MacLean, D.C., Roark, O.F., Folkerts, G., & Schneider, R.E. (1969). Influence of mineral nutrition as the sensitivity of tomato plants to hydrogen fluoride. *Env. Sci. Tech.*, **3**, 1201–1204.

Magel, E., Holl, W., & Ziegler, H. (1988). The status of reserve carbohydrates and adenine nucleotides in needles of spruce trees (*Picea abies* L. Karst.) after treatment with ozone and acid with independence on the soil type. (Abst.) In *Int. Symp. on Plants and Pollutants in Developing Countries* (Ed. M. Ozturk) p. 93. Izmir, Turkey, 1988. 134 pp.

Malhotra, S.S., & Blauel, R.A. (1980). *Diagnosis of Air Pollutant and Natural Stress Symptoms on Forest Vegetation in Western Canada*, Canadian For. Serv. Info. Rept. NOR-X-228, Edmonton, 84 pp.

Malhotra, S.S., & Hocking, D. (1976). Biochemical and cytological effects of sulphur dioxide on plant metabolism. *New Phytologist*, **76**, 227–237.

Malhotra, S.S., & Khan, A.A. (1984). Biochemical and physiological impact of major pollutants. In M. Treshow (ed.), *Air Pollution and Plant Life*, Wiley, Chichester, 486 pp.

Mandl, R.H., Weinstein, L.H., McCune, D.C., & Keveny, M. (1973). A cylindrical, open-top chamber for the exposure of plants to air pollutants in the field. *J. Environ. Qual.*, **2**, 371–376.

Manion, P.D. (1981). *Tree Disease Concepts*, Prentice-Hall, New York, 399 pp.

Manning, W.J. (1975). Interactions between air pollutants and fungal, bacterial and viral plant pathogens. *Environ. Pollut.*, **9**, 87–90.

Manning, W.J. (1978). Chronic foliar injury: effects on plant root development and possible consequences. *Calif. Air Environ.* 7, 3–4.

Manning, W.J., Feder, W.A., Perkins, I., & Glickiman, M. (1969). Ozone injury and infection of potato leaves by *Botrytis cinerea*. *Plant Dis. Rep.*, **53**, 691–693.

Manning, W.J., Feder, W.A., & Vardano, P.M. (1974). Suppression of oxidant injury to benomyl: Effects on yields of bean cultivars in the field. *J. Environ. Quality*, **3**, 1–3.

Mansfield, T.A. (1973). The role of stomata in determining the response of plants to air pollutants. *Current Adv. Plant Sci.*, **2**, 11–20.

Manual, E.H., Horst, R.L., Brennau, K.M., Lanen, W.N., Duff, M.C., & Tapiero, J.K. (1981). *Benefits analyses of alternate secondary national ambient air quality standards for sulfur dioxide and total suspended particulates*, Vol. IV, EPA 68-01-3392 0AQPS, US EPA, Durham, NC.

Masuch, G., Kettrup, A., Mallant, R.K., & Slanina, J. (1985). Histological effects of H$_2$O$_2$ on the structure of beech leaves and spruce needles. In *International Workshop on Physiology and Biochemistry of Stressed Plants*, pp. 14–24, Neuherberg, FRG, May 20–21.

Materna, J. (1972). Einfluss niedriger Schwelfedroxyd-Kohzentrationen auf der Fichte. *Mitt. Forst. Bundes Anstalt.*, **97**, 219–232.

Materna, J. (1984). Impact of air pollution in natural ecosystems. In M. Treshow (ed.), *Air Pollution and Plant Life*, pp. 397–416, Wiley, Chichester.

Matsushima, S., & Brewer, R.F. (1972). Influence of sulfur dioxide and hydrogen fluoride as a mix or reciprocal exposure on citrus growth and development. *J. Air Poll. Contr. Assoc.*, **22**, 710–713.

Mayrhofer, J. (1893). Über Pflanzen beschädigung, veranlosst durch den Betrieb einer Superphosphate fabrik. *Freie Ver. Bayer Vertreter Angew. Chemie Ber.*, **10**, 127–129.

McClenahen, J.R. (1978). Community changes in a deciduous forest exposed to air pollution. *Can. J. For. Res.*, **8**, 232–438.

McCool, P.M., Menge, J.A., & Taylor, O.C. (1979). Effects of ozone and HCl gas on development of the mycorrhiza fungus *Glomas fasciculatus* and growth of Troyer citrang. *J. Am. Soc. Hort. Sci.*, **104**, 151–154.

McCune, D.C. (1969). *On Establishment of Air Quality Criteria, with Reference to the Effects of Atmospheric Fluoride on Vegetation*, Air Quality Monograph 69-3, American Petroleum Institute. New York, 33 pp.

McCune, D.C. (1983). Terrestrial vegetation air pollution interactions: gaseous pollutants—hydrogen fluoride and sulphur dioxide. In *Air Pollutions and their Effects on the Terrestrial Ecosystems* (Eds. S.V. Krupa & A.H. Legge). Proc. International Conference on Current Status and Future Needs of Research on Effects of Technology, Banff, Alberta. Wiley, Chichester.

McCune, D.C., Hitchcock, A.E., Jacobson, J.S., & Weinstein, L.H. (1965). Fluoride accumulation and growth of plants exposed to particulate cryolite in the atmosphere. *Contrib. Boyce Thompson Inst. Plant Res.*, **23**, 1–12.

McCune, D.C., Hitchcock, A.E., & Weinstein, L.H. (1966). Effect of mineral nutrition on the growth and sensitivity of gladiolus to hydrogen fluoride. *Contr. Boyce Thompson Inst. Plant. Res.*, **23**, 295–299.

McCune, D.C., Weinstein, L.H., Mancini, J.F., & van Leuken, P. (1973). Effects of hydrogen fluoride on plant pathogen interactions. *Proc. Int. Clean Air Conf.*, Dusseldorf.

McIlveen, W.D., Spots, R.A., & Davis, D.D. (1975). The influence of soil zinc on nodulation, mycorrhizae and ozone sensitivity of pinto beans. *Phytopathology*, **65**, 645–647.

McLaughlin, S.B. (1985). Effects of air pollutants on forests: A critical review. *J. Air Poll. Contr. Assoc.*, **35**, 512–534.

McLaughlin, S.B., & Taylor, G.E., Jr. (1981). Relative humidity: Important modifier of pollutant uptake by plants. *Science*, **211**, 167–169.

Menser, H.A., Jr (1963). The effect of ozone and controlled environment factors on four varieties of tobacco *Nicotiana tabacum* L. PhD thesis, University of Maryland, College Park, MD.

Menser, H.A., & Heggestad, H.E. (1966). Ozone and sulfur dioxide synergism: Injury to tobacco plants. *Science*, **153**, 424–425.

Menser, H.A., & Street, O.E. (1962). Effects of air pollution, nitrogen levels, supplemental irrigation, and plant spacing on weather fleck and leaf losses in Maryland tobacco. *Tobacco Sci.*, **6**, 167–171.

Middleton, J.T. (1956). Response of plants to air pollution. *J. Air Poll. Contr. Assoc.*, **6**, 7–9.

Middleton, J.T., Kendrick, J.B., Jr, & Schwalm, H.W. (1950). Injury to herbaceous plants by smog or air pollution. *Plants Dis. Rep.*, **34**, 245–252.

Middleton, J.T., Kendrick, J.B., Jr, & Darley, E.F. (1953). Air pollution injury to crops. *Calif. Agric.*, **7**, 11–12.

Middleton, J.T., Kendrick, J.B., Jr, & Darley, E.F. (1955). Air-borne oxidants as plant-damaging agents. *Proc. 3rd Nat. Air Poll. Symp.*, 191–198.

Middleton, J.T., Darley, E.F., & Brewer, R.E. (1957). Damage to vegetation from polluted atmospheres. *Proc. Am. Petrol. Inst.*, Sect. III, 8 pp.

Middleton, J.T., Darley, E.F., & Brewer, R.F. (1958). Damage to vegetation from polluted atmospheres. *J. Air Poll. Contr. Assoc.*, **8**, 9–15.

Miller, C.A., & Davis, D.O. (1981). Effect of temperature on stomatal conductance and ozone injury of pinto bean leaves. *Plant Dis.*, **65**, 750–751.

Miller, J.E., & Miller, G.W. (1974). Effect of fluoride on mitochondrial activity in higher plants. *Physiol. Plant.*, **32**, 115–121.

Miller, P.R. (1973a). Susceptibility to ozone of selected western conifers. *2nd Int. Congr. Plant Pathol.*, Abstract'No. 0579.

Miller, P.R. (1973b). Oxidant-induced community change in a mixed conifer forest. In J. Naegel (ed.), *Air Pollution Damage to Vegetation*, pp. 101–117, Adv. Chem. Serv. No. 122, Am. Chem. Soc., Washington, DC.

Miller, P.R., & Elderman, M.J. (1977). Photochemical oxidant air pollution effects on a mixed conifer ecosystem. *Ecological Research Series*, EPA 600/3-77-104.

Miller, P.R., & Evans, L.S. (1974). Histopathology of oxidant injury and winter fleck injury on needles of western pines. *Phytopathology*, **64**, 801–806.

Miller, P.R., & McBride, J.R. (1975). Effects of air pollutants on forests. In J.B. Mudd & T.T. Kozlowski (eds), *Responses of Plants to Air Pollutants*, pp. 196–236, Academic Press, New York.

Miller, V.L., Johnson, F., & Allmendinger, D.F. (1948). Fluorine analysis of Italian prune foliage affected by marginal scorch. *Phytopathology*, **38**, 30–37.

Miller, P.R., Parmeter, J.R., Flick, B.H., & Martinez, C.W. (1969). Ozone dose response of ponderosa pine seedlings. *J. Air Poll. Contr. Assoc.*, **19**, 435–438.

Mjelde, J.W., Adams, R.M., Dixon, B.L., & Garcia, P. (1984). Using farmers actions to measure crop loss due to air pollution. *J. Air Poll. Contr. Assoc.*, **32**, 360–364.

Mooney, H.A., Vitousek, P.M., & Matson, P.A. (1987). Exchange of materials between terrestrial ecosystems and the atmosphere. *Science*, **233**, 926–932.

Moskowitz, P.D., Coveney, E.F., Mederios, W.H., & Morris, S.C. (1982). Oxidant air pollution: A model for estimating effects on U.S. vegetation. *J. Air Poll. Contr. Assoc.*, **32**, 155–160.

Mudd, J.B. (1963). Enzyme inactivation by peroxyacetyl nitrate. *Arch. Biochem. Biophys.* **102**, 59–65.

Mudd, J.B. (1975). Peroxacyl nitrates. In J.B. Mudd & T.T. Kozlowski (eds), *Responses of Plants to Air Pollution*, pp. 97–119, Academic Press, New York.

Mudd, J.B., Banerjee, S.K., Dooley, M.M., & Knight, K.L. (1984). Pollutants and plant cells; effects on membranes. In *Gaseous Air Pollutants and Plant Metabolism* (Eds. M.J. Koziol & F.R. Whatley), pp. 105–116. Butterworths, London, xiv + 466 pp., illus.

Mudd, J.B., & McManus, T.T. (1969). Products of the reaction of peroxyacetyl nitrate with sulfhydryl compounds. *Arch. Biochem. Biophys.*, **132**, 237–241.

Muir, P.S., & McCune, B. (1988). Lichens, tree growth, and foliar symptoms of air pollution: Are the stories consistent? *J. Environ. Qual.*, **17**, 361–370.

Mukerji, S.K., & Yang, S.F. (1974). Phosphoenolpyruvate carboxylase from spinach leaf tissue: Inhibition by sulfite ion. *Plant Physiol.*, **53**, 829–834.

Munch, E. (1933). Winter schäden an immergrünen Geholzen. *Ber. Deutsch Bot. Ges.*, **51**, 21.

Musselman, R.C., & Sterrett, J.L. (1988). Sensitivity of plants to acidic fog. *J. Environ. Qual.*, **17**, 329–333.

Nakamura, H., & Saka, H. (1978). Photochemical oxidants injury in rice plants. III. Effect of ozone on physiological activities in rice plants. *Japan. J. Crop Sci.*, **47**, 707–714.

Nash, T.H., III, & Sigal, L.L. (1980). Sensitivity of lichens to air pollution with an emphasis on oxidant air pollutants. In *Proc. Symp. on Effects of Air Pollutants on Mediterranean and Temperate Forest Ecosystems*, pp. 117–124, Riverside, California.

National Academy of Sciences (1971). Effects of fluoride on vegetation. In *Fluorides* (Committee on Biological Effects of Atmospheric Pollutants), pp. 77–132, NAS, Washington, DC.

National Academy of Sciences (1977a) *Arsenic* (Committee on Medical and Biological Effects of Environmental Pollutants), NAS, Washington, DC, 322 pp.

National Academy of Sciences (1977b). *Ozone and other Photochemical Oxidants* (Committee on Medical and Biological Effects of Environmental Pollutants), NAS, Washington, DC, 719 pp.

National Acid Precipitation Assessment Program (1986). *Annual Report, 1986*, Washington, DC, 163 pp.

National Acid Precipitation Assessment Program (1987). *Interim Assessment, the Causes and Effects of Acid Deposition*, US Govt. Printing Office, Washington, DC.

National Research Council (1983). *Changing Climate*. A conservative report by the NRC carbon dioxide assessment committee. National Academy Press, Washington, DC.

National Research Council of Canada (1939). *Effect of Sulphur Dioxide on Vegetation*. NRC of Canada, ix + 447 pp., illus.

Nihlgard, B. (1985). The ammonium hypothesis: An additional explanation to the forest dieback in Europe. *Ambio*, **14**, 2–8.

Nilsson, S., & Duinker, P. (1987). The extent of forest decline in Europe. *Environment*, **29**, 4–10, 30–31.

Nobel, P.S., & Wang, C. (1973). Ozone increases the permeability of isolated pea chloroplasts. *Arch. Biochem. Biophys.*, **157**, 388–394.

Noble, W. (1965). Smog damage to plants. *Lasca Leaves*, **15**, 24.

Oden, S. (1976). The acidity problem: An outline of concepts. In L.S. Dochinger & T.A. Seliga (eds), *Proc. 1st Inter. Symp. on Acid Precip. and the Forest Ecosystem*, pp. 1–42, NE Forest Exp. Sta. Rep. NE-23.

Oden, S., & Ahl, T. (1970). The acidification of Scandinavian lakes and rivers. *YMER ÅRSBOK*, 103–122 (Swedish with English summary).

OECD (1978). *Plant Damage Caused by SO_2*. Report of the workshop held at the OECD, Paris 7–8 June, ENV/AIR/78, **15**, 1–30.

Ogata, G., & Maas, E.V. (1973). Interactive effects of salinity and ozone on growth and yield of garden beet. *J. Environ. Qual.*, **2**, 518–520.

Okano, L., Ito, O., Takeba, G., Shimizu, A., & Totsuka, T. (1984). Alteration of ^{13}C-assimilate partitioning in plants of *Phaseolus vulgaris* exposed to ozone. *New Phytol.* **97**, 155–163.

Ormrod, D.P. (1977). Cadmium and nickel effects on growth and ozone sensitivity of pea. *Water Air Soil Pollut.*, **8**, 263–270.

Ormrod, D.P. (1982). Air pollution interactions in mixtures. In *Effects of Gaseous Air Pollution in Agriculture and Horticulture* (Eds. M.H. Unsworth & D.P.

Ormrod), pp. 207–213. Butterworths, London. xiv + 532 pp. illus.

Ormrod, D.P. (1984). Impact of trace element pollution on plants. In M. Treshow (ed.), *Air Pollution and Plant Life*, pp. 291–319, Wiley, Chichester.

Oshima, P.J. (1978). The impact of sulfur dioxide on vegetation: A sulfur dioxide–ozone response model. Calif. Air Resource Board, Agreement No. A6-162-30.

Oshima, R.J., & Bennett, J.P. (1978). Experimental design and analysis. In W.W. Heck, S.V. Krupa & S.N. Linzon (eds), *Methodology for the Assessment of Air Pollution Effects on Vegetation*, Ch. 4, pp. 1–23, Air Pollution Control Association, Pittsburgh.

Oshima, R.J., Poe, M., Braegelmann, P.K., Baldwin, D.W., & Way, V. von (1976). Ozone dosage–drop loss function for alfalfa: A standard method for assessing crop losses from ar pollution. *J. Air Poll. Contr. Assoc.*, **26**, 861–865.

Otto, H.W., & Daines, R.H. (1969). Plant injury by air pollutants: Influence of humidity on stomatal aperature and plant response to ozone. *Science*, **163**, 1209–1210.

Overrein, L., Seip, H.M., & Tollan, A. (1981). *Acid Precipitation: Effects on Forest and Fish*. Final report of the SNSF, project 1972–1980, Norwegian Ministry of the Environment, Oslo, Norway.

Pack, M.R. (1966). Response of tomato fruiting to hydrogen fluoride as influenced by calcium nutrition. *J. Air Poll. Contr. Assoc.*, **16**, 541–544.

Pack, M.R., Hill, A.C., Thomas, M.D., & Transtrum, L.G. (1959). Determination of gaseous and particulate inorganic fluorides in the atmosphere. Symp. on air pollution control, *Spec. Tech. ASTM Publ. No. 281*, pp. 27–44.

Padgett, J., & Richmond, H. (1983). The process of establishing and revising national ambient air quality standards. *J. Air Pollut. Contr. Assoc.*, **33**, 13–16.

Page, A.L., Bingham, F.T., & Chang, A.C. (1981). Cadmium. In N.W. Lepp (ed.), *Effect of Heavy Metal Pollution on Plants*, pp. 77–109, Applied Science, London.

Page, W.P., Arbogast, G., Fabian, R., & Clecka, J. (1982). Estimation of economic losses to the agricultural sector from airborne residuals in the Ohio River Basin region. *J. Air Poll. Contr. Assoc.*, **32**, 151–154.

Pandya, N., & Bedi, S.J. (1988). Growth and yield reduction in egg plants at high air pollution zones. *Int. Symp. on plants and pollutants in developed and developing countries* (Abstract), 22–28 August, Izmir, Turkey.

Parmeter, J.R., Bega, R.V., & Neff, R. (1962). A chlorotic decline of ponderosa pine in southern California. *Plant Dis. Rep.*, **46**, 269–273.

Parmeter, J.R., & Miller, P.R. (1968). Studies relating to the cause of decline and death of ponderosa pine in southern California. *Plant Dis. Rep.* **52**, 707–711.

Pell, E.J., & Brennan, E. (1973). Changes in respiration, photosynthesis, adenosine 5-triphosphate, and total adenylate content of ozonated pinto bean foliage as they relate to symptom expression. *Plant Physiol.*, **51**, 378–381.

Pell, E.J., & Pearson, N.S. (1983). Ozone induced reduction in quantity of ribulose-1,5-biphosphate carboxylase in alfalfa foliage. *Plant Physiol.*, **73**, 185–187.

Pell, E.J., & Weissberger, W.C. (1976). Histopathological characterization of ozone injuries to soybean foliage. *Phytopathology*, **66**, 856–861.

Pell, E.J., Arny, C.I., & Pearson, N.S. (1987). Impact of simulated acid precipitation on quantity and quality of a field grown potato crop. *Environ. Exp. Bot.*, **27**, 6–14.

Perkins, D.F., Millar, R.O., & Neep, P.E. (1980). Accumulation of airborne fluoride by lichens in the vicinity of an aluminum reduction plant. *Environ. Pollut.*, **21**, 155–168.

Peterson, D.L., Wakefield, V.A., & Arbaugh, M.J. (1987). Detecting the effect of ozone pollution on growth of Jeffrey pine in the Sierra Nevada, Calif. In *Proc. Internat. Symp. Ecol. Aspects of Tree Ring Analysis*, pp. 401–409.

Pinkerton, J.E., & Lefohn, A.S. (1987). The characterization of ozone data for sites located in forested areas of the eastern United States. *J. Air Poll. Contr. Assoc.*, **37**, 1005–1010.

Poovaiah, B.W., & Wiebe, H.H. (1973). Influence of hydrogen fluoride fumigation on the water economy of soybean plants. *Plant Physiol.*, **51**, 396–399.

Posthumus, A.C. (1985). The effects of ozone and other photochemical oxidants on vegetation. In WHO Air Quality Guidelines, *Ecological Effects of Air Pollutants*, Summary Report on ecological effects of air pollutants, Neukirchen, Austria, June 24–28.

Preston, E.M. (1979). The ecological implications of sulfur dioxide exposure for native grasslands. *72nd Ann. Air Poll. Contr. Assoc.*, Cincinnatti, OH.

Prinn, R., Cunnold, D., Rasmussen, R., Simmonds, P., Alyea, F., Crawford, A., Fraser, P., & Rosen, R. (1987). Atmospheric trends in methylchloroform and the global average for the hydroxyl radical. *Science*, **238**, 945–950.

Prinz, B. (1983). Gedanken zum Stand der Diskussion über die Ursache der Waldschäden in der Bundesrepublik Deutschland. *Der Forst- und Holzwirt.*, **38**, 460–467.

Prinz, B. (1985). Effects of air pollutions on forests: Critical review discussion papers. *J. Air Poll. Contr. Assoc.*, **35**, 913–924.

Prinz, B. (1987). Causes of forest damage in Europe. *Environment*, **29**, 11–15, 32–37.

Prinz, B., Krause, H.M., & Stratmann, H. (1982). *Vorläufige Bericht der Landesanstalt für Immissionsschutz über Untersuchungen zur Aufklärung der Waldschäden in der Bundesrepublik Deutschland*, LIS-Bericht Nr. **28**, Landesanstalt für Immissionsschutz des Landes NW, Essen, 54 pp.

Prinz, B., Krause, B.H.M., & Jung, K.D. (1984). Neuere Untersuchungen der Lis zu den neuartigen Waldschäden. *Dusseldorfer Geobotanische Kolloquien*, **1**, 25–36.

Psenak, M., Miller, G.W., Yu, M.H., & Lovelace, C.J. (1977). Separation of malic dehydrogenase tissue in relation to fluoride treatment. *Fluoride*, **10**, 63–72.

Ramanathan, V. (1988). The greenhouse theory of climate change: A test by an inadvertent global experiment. *Science*, **240**, 293–299.

Rasmussen, R.A., & Khalil, M.A.K. (1986). Atmospheric trace gases: Trends and distributions over the last decade. *Science*, **232**, 1623–1624.

Ratsch, H.C. (1974). *Heavy-metal accumulation in soil and vegetation from smelter emissions*, US Environmental Research Center Publ. EPA-660/3-74-012, US Supt. Documents, Washington, DC, 23 pp.

Rawlings, J.O., Lesser, V.M., Heagle, A.S., & Heck, W.W. (1988). Alternative dose metrics to characterizing ozone impact on crop yield loss. *J. Environ. Qual.*, **17**, 285–291.

Ray, S. (1988). Legislative control of antropogenic environmental pollution: Scope, problems and possibilities. *Int. Symp. on plants and pollutants in developed and developing countries* (Abstract) 22–28 August, Izmir, Turkey.

Rehfuess, K.E. (1981). On the impact of acid precipitation on forested ecosystems. *Forstwiss. Centralbl.*, **100**, 363–381.

Rehfuess, K.E. (1983). Walderkrankungen und Immissionsens eine Zwischenbilanz. *Allg. Forstzeitschn.*, **38**, 601–610.

Rehfuess, K.E. (1985). On the causes of decline of Norway spruce (*Picea abies* Karst) in central Europe. *Soil Use and Management*, **1**, 30–32.

Rehfuess, K.E., & Rodenkirken, H. (1984). Über die Nadelrote-Erkrankung der

Fichte (*Picea abies* Karst) in Suddeutschland. *Forstwissens. Centralblatt.*, **103**, 249–262.

Reich, P.B., & Schoettle, A.W. (1987). Effects of ozone and acid rain on white pine (*Pinus strobus*) seedlings grown in five soils. I. Net photosynthesis and growth. *Can. J. Bot.*, **65**, 977–987.

Reinert, R.A., & Weber, D.E. (1980). Ozone and sulfur dioxide-induced changes in soybean growth. *Phytopathology*, **70**, 914–916.

Reinert, R.A., Tingey, O.T., Heck, W.W., & Wickliff, C. (1969). Tobacco growth influenced by low concentrations of sulfur dioxide and ozone. *Agron. Abstr.*, **61**, 34.

Reinert, R.A., Heagle, A.S., & Heck, W.W. (1975). Plant responses to pollutant combinations. In J.B. Mudd & T.T. Kozlowski (eds), *Responses of Plants to Air Pollution*, pp. 159–177, Academic Press, New York.

Rich, S., & Tomlinson, H. (1968). Effects of ozone on conidiophores and conidia of *Alternaria solani*. *Phytopathology*, **58**, 444–446.

Rich, S., & Turner, N.C. (1972). Importance of moisture on stomatal behavior of plants subjected to ozone. *J. Air Pollut. Contr. Assoc.*, **22**, 718–721.

Richards, B.L., Middleton, J.T., & Hewitt, H.B. (1958). Air pollution with relation to agronomic crops. V. Oxidant stipple of grape. *Agron. J.*, **50**, 559–561.

Richards, B.L., Taylor, O.C., & Edmunds, G.F., Jr. (1968). Ozone needle mottle of pine in Southern California. *J. Air Poll. Contr. Assoc.*, **18**, 73–77.

Richardson, D.H.S. (1975). *The Vanishing Lichens*, David & Charles, Newton Abbot, UK, 231 pp.

Ridley, J.D., & Sims, E.T., Jr (1967). The responses of peaches to ozone during storage *SC Agr. Exp. Sta. Tech. Bull.*, **1**, 1–24.

Roberts, J.J. (1984). Air quality management. In S. Calvert & H.M. Englund (eds), *Handbook of Air Pollution Technology*, pp. 969–1038, Wiley, New York.

Robock, A. (1983). Global mean sea level: Indicator of climatic change? [Letter in response to an article by R. Ethkins & E. Epstein, *Science*, **215**, 287 (1982)]. *Science*, **219**, 996–998.

Roelofs, J.G.M., Kempers, A.J., Houdijk, A.L.F.M., & Jansen, J. (1985). The effect of airborne ammonium sulphate on *Pinus nigra* var. *maritima* in The Netherlands. *Pl. Soil*, **84**, 45–56.

Roelofs, J.G.M., Boxman, A.W., & van Dijk, H.F.G. (1987). Effects of airborne ammonium on natural vegetation and forests. In W.A.H. Asman & S.M.A. Diederen (eds), *Ammonia and Acidification Proceedings*, Symposium of the European Association for the Science of Air Pollution held at the National Institute of Public Health & Environmental Hygiene, Bilthoven, Netherlands, 13–15 April.

Roholm, K. (1937). *Fluorine Intoxication Hgt.*, Nordisk Forlag., Arnold Busch, Copenhagen, 364 pp.

Rotty, R.M., & Reister, O.B. (1986). Use of energy scenarios in addressing the CO_2 question. *J. Air Pollut. Contr. Assoc.*, **36**, 1111–1115.

Rowe, R.D., & Miller, C. (1985). *Economic Assessment of the Effects of Air Pollution on Agricultural Crops in the San Joaquin Valley*, Energy & Resource Consultants, Boulder, CO, 183 pp.

Runeckles, V.C. (1984). Impact of air pollutant combinations on plants. In M. Treshow (ed.), *Air Pollution and Plant Life*, pp. 239–258, Wiley, Chichester.

Runeckles, V.C., & Palmer, K. (1987). Pretreatment with nitrogen dioxide modifies plant response to ozone. *Atmos. Environ.*, **21**, 717–719.

Runeckles, V.C., & Resh, H.M. (1975). The assessment of chronic ozone injury to leaves by reflection spectrophotometry. *Atmos. Environ.*, **9**, 447–452.

Runeckles, V.C., Palmer, K., & Giles, K. (1978). Effects of sequential exposures to NO_2 and O_3 on plants. *3rd. Int. Congr. Plant Pathology*, Munich (Abstract), 343 pp.

Russell, E.W. (1973). *Soil Conditions and Plant Growth*, 10th edn, Longman, New York.

Ryan, J.W. (1981). *An Estimate of the Non-health Benefits of Meeting the National Ambient Air Quality Standards*, SRI Report to the National Commission on Air Quality, Menlo Park, CA.

Sakal, R.R., & Rohlf, F.J. (1969). *Biometry*, Freeman, San Francisco.

Saunders, P.J.W. (1975) Air pollutants, microorganisms and interaction phenomena. *Environ. Pollut.*, **9**, 85–86.

Scheffer, T.C., & Hedgecock, G.G. (1955). Injury to northwestern forest trees by sulfur dioxide from smelters. *USDA Tech. Bull. 1117.*, 49 pp.

Scheutte, L.R. (1971). Responses of the primary infection process of *Erisiphe graminis* f. sp. *hordei* to ozone. PhD thesis, University of Utah, Salt Lake City, 77 pp.

Schmitt, G. (1986). The temporal distribution of trace element concentrations in fog water during individual fog events. In H.W. Georgii (ed.), *Atmospheric Pollutants in Forest Areas*, pp. 129–141, Reidel, Dordrecht.

Schopfer, W. Von, & Hradetzky, J. (1986). Zuwachstruckgang in erkrankten Fichtenan und Tannenbestanden auswertungsmethoden und Ergebnisse. *Forstwissens. Centralblatt.*, **6**, 446–470.

Schreiber, V., Vidaver, W., Runeckes, V.C., & Rosen, P. (1978). Chlorophyll fluorescence assay for ozone injury in intact plants. *Plant Physiol.*, **61**, 81–84.

Schrenk, von H., & Spaulding, P. (1909). Disease of deciduous forest trees. USDA *Bu. Plant Industry Bull.*, No. 149, 85 pp.

Schroeder, von J., & Ruess, C. (1883). *Die Beschädegung der Vegetation durch Rauch and die Oberharzer Hüttenrauchschäden*. Parey, Berlin.

Schütt, P. (1983). Botanische Aspekte der Forschung zum Waldsterben: Ergebnisse und Perspektiven. In *Tagungsber. Symp. Saurer Regen–Waldschäden*, January, pp. 36–37.

Schütt, P., & Cowling, E.B. (1985). Waldsterben, a general decline of forests in central Europe: Symptoms, development and possible causes. *Plant Dis.*, **69**, 548–558.

Schütt, P., Koch, W. Blaschke, H., Lang, K.J., Schusk, H.J., & Simmerer, S. (1983). *So stirbt der Wald – Schadbilder und Krankheitsverlauf*. BLV-Verlagsgesellschaft, München, 127 pp.

Schweingruber, F.H. (1986). Abrupt growth changes in conifers. *IAWA Bull.*, **7**, 277–283.

Seidel, S., & Keyes, D. (1983), *Can We Delay a Greenhouse Warming?* Environmental Protection Agency, Washington, DC.

Seufert, G., & Arndt, V. (1986). Unterschungern zum Stoffhaushalt in definiert belasteten Model Lokosystemen mit jungen Waldbaumen in Wirkungen von Luftverunneinigungen auf Waldbaume und Waldboden. *Allg. Forst. Z.*, **1**, 2–13.

Sherzer, A.J., & McClenahen, J.R. (1989). Effects of ozone or sulfur dioxide on pitch pine seedlings. *J. Env. Qual.*, **18**, 57–61.

Shimazaki, K., & Sugahari, K. (1980). Inhibition site in electron transport system in chloroplasts by fumigation of lettuce leaves with SO_2. *Studies on the Effects of Air Pollutants on Plants and Mechanisms of Phytotoxicity: Res. Rep. Natl. Inst. Environ. Stud. Japan*, **11**, 79–89.

Shriner, D.S., Abner, C.H., & Mann, L.K. (1977). Rainfall simulation for environmental application. ORNL-5151, Oak Ridge National Laboratory Report, Oak Ridge, TN.

Shriner, D.S., Cure, W.W., Heagle, A.S., Heck, W.W., Johnson, S.W., Olson, R.J., & Kelley, J.M. (1982). *On Analysis of Potential Agriculture and Forestry Impacts of Long-term Transport of Air Pollutants.* ORNL-590, Oak Ridge National Laboratory, Oak Ridge, TN.

Sidhu, S.S. (1980). Patterns of fluoride accumulation in boreal forest species under perennial exposure to emissions from a phosphorus plant. *Atmospheric Pollution, 1980. Studies on Environmental Science*, **8**, 425–432.

Sidhu, S.S., & Staniforth, S.S. (1986). Effects of atmospheric fluoride on foliage, cone and seed production of balsam fir, black spruce and larch. *Can. J. Bot.*, **64**, 923–931.

Skarby, L., Troeng, E., & Bostrom, C.A. (1987). Ozone uptake and effects of transpiration, net photosynthesis, and dark respiration in Scots pine. *For. Sci.*, **33**, 801–808.

Skeffington, R.A.B., & Roberts, T.M. (1985). The effects of ozone and acid mist on Scots pine seedlings. *Oecologia*, **65**, 201–206.

Skelly, J.M., Yang, Y.S., Chevone, B.I., Long, S.J., Nellessen, J.E., & Winner, W.E. (1983). Ozone concentrations and their influence on forest species in the Blue Ridge Mountains of Virginia. In D.D. Davis, A.A. Miller, & L. Dochinger (eds), *Air Pollution and the Productivity of the Forest*, pp. 143–159, Isaac Walton League, Washington, DC.

Smith, A.L. (1921). *Lichens*, Cambridge University Press, London, 464 pp.

Smith, G., Greenhalgh, B., Brennan, E., & Justin, J. (1987). Soybean yield in New Jersey selective to ozone pollution and antioxidant application. *Plant Disease*, **71**, 121–125.

Smith, M., & Brown, D. (1981). *Estimating Economic Loss due to Ozone Damage to Corn, Wheat and Soybeans.* Ann. Mtg. Amer. Agric. Economics Assoc., Logan, UT.

Smith, W.H. (1984). Pollutant uptake by plants. In M. Treshow (ed.), *Air Pollution and Plant Life*, pp. 417–450, Wiley, Chichester.

Soderlund, R. (1981). Dry and wet deposition of nitrogen compounds. In F.E. Clark & T. Rosswall *Terrestrial Nitrogen Cycles. Ecological Bulletin*, pp. 123–130, Swedish Natural Science Research Council, Stockholm.

Soikkeli, S. (1981). Comparison of cytological injuries in conifer needles from several polluted industrial environments in Finland. *An. Bot. Fenn.*, **18**, 47–61.

Soikkeli, S., & Karenlampi, L. (1984). Cellular and ultrastructural effects. In M. Treshow (ed.), *Air Pollution and Plant Life*, pp. 159–174. Wiley, Chichester.

Sokal, R.R., & Rohlf, J. (1969). *Biometry.* Freeman, San Francisco. xxi + 776 pp., illus.

Sorauer, P. (1886). *Handbuch der Pflanzenkrankheiten*, 2nd edn (3rd. edn 1908–1913), Parey, Berlin.

Sorauer, P. (1914). Non-parasitic diseases. In *Manual of Plant Diseases*, Vol. 1, Record Press, Wilkes Barre, PA.

Spalding, D.H. (1968). Effects of ozone atmospheres on spoilage of fruits and vegetables. *US Dept. Agric. Marketing Res. Rept.*, **801**.

Spikes, J.D., Lumry, R.L., & Rieske, J.S. (1955). Inhibition of the photochemical activity of chloroplasts. I. Salts. *Arch. Biochem. Biophysics*, **55**, 25–37.

Sprugel, D.G., Miller, J.E., Muller, R.N., Smith, H.J., & Xerikos, P.B. (1980). Sulfur dioxide effects on yield and seed quality in field-grown soybeans. *Phytopathology*, **70** 1129–1133.

Stanford Research Institute (1981). *An estimate of non-health benefits of meeting the secondary national ambient Air Quality Standards*, Prepared for the National Commission on Air Quality, Washington, DC.

Stasiuk, W.E., & Coffrey, P.N. (1974). Rural and urban ozone concentrations in New York State. *J. Air Poll. Contr. Assoc.*, **24**, 564–568.

Stauffer, B., Fischer, G., Neftel, A., & Oeschger, H. (1985). Increase of atmospheric methane recorded in Antarctic ice core. *Science*, **229**, 1386–1388.

Stephens, E.R., Hanst, P.L., Doerr, R.C., & Scott, W.E. (1956). Reactions of NO_2 and organic compounds in air. *Ind. Eng. Chem.*, **48**, 1498–1504.

Stephens, E.R., Darley, E.C., Taylor, O.C., & Scott, W.E. (1961). Photochemical reaction products in air pollution. *Int. J. Air Poll.*, **4**, 79–100.

Stern, A.C., Boubel, W., Turner, D.B., & Fox, D.L. (1984). *Fundamentals of Air Pollution*. 2nd ed, Academic Press, Orlando, FL, 530 pp.

Stöckhardt, J.A. (1853). Untersuchung junges Fichten und Kierfern, welche durch den Rauch des Antons-Hutte Krank geworden sind. *Thar. Forstl. Shb.*, **9**, 169–172.

Stöckhardt, J.A. (1871). Untersuchungen über die Schadlicken Erwirkungen des Hutten- und Steinkohlenrauches auf das Wachstum der Pflanzen, insbesondere der Fichte und Tanne. *Tharander Forstl.*, **21**, 218–230.

Stoklasa, J. (1923). *Die Beschädigung der Vegetation durch Rauchgase und Fabriksexhalation*, Urban & Schwarzenberg, Munich.

Stolarski, R.S. (1988). The Antarctic ozone hole. *Scientific American*, **258**, 30–36.

Stolte, K.W. (1982). The effects of ozone on chaparral plants in the California South Coast Air Basin. MS thesis, University of California, Riverside, CA, 100 pp.

Sulzbach, C.W., & Pack, M.R. (1972). Effects of fluoride on pollen germination, pollen tube growth and fruit development in tomato and cucumber. *Phytopathology*, **62**, 1247–1253.

Swanson, E.S, Thomson, W.W., & Mudd, J.B. (1973). The effect of ozone on bean leaf cell membranes. *Can. J. Bot.*, **51**, 1213–1219.

Tamm, C.O., & Cowling, E.B. (1976). Acidic precipitation and forest vegetation. In L.S. Dochinger & T.A. Seliga (eds), *Proc. 1st Internat. Symp. on Acid Precip. and the Forest Ecosystem*, pp. 846–855. NE Forest Exp. Sta. Rep. NE-23.

Taylor, O.C. (1969). Importance of peroxyacetyl nitrate (PAN) as a phytotoxic air pollutant. *J. Air Poll. Contr. Assoc.*, **19**, 347–351.

Taylor, O.C. (1973). *Oxidant Air Pollution Effects on a Western Coniferous Forest Ecosystem*, Task C. Report, University of California Statewide Air Pollution Research Center, Riverside.

Taylor, O.C. (1974). Air pollution effects influenced by plant-environment interaction. In W.M. Duggar (ed.), *Air Pollution Effects on Plant Growth*, pp. 1–7, A. Chem. Soc. Symp. Ser. 3, Washington, DC.

Taylor, O.C. (1984). Organismal responses of higher plants to atmospheric pollutants: Photochemical and other. In M. Treshow (ed.), *Air Pollution and Plant Life*, pp. 215–230. Wiley, Chichester.

Taylor, O.C., & MacLean, D.C. (1970). Nitrogen oxides and peroxyacyl nitrates. Section E1-14. In: *Recognition of Air Pollution Injury to Plants. A Pictorial Atlas*. Air Poll. Contr. Assoc., Inf. Rept. No. 1.

Taylor, G.S., & Rich, S. (1974). Ozone injury to tobacco in the field influenced by soil treatments with Benomyl and Carboxin. *Phytopathology*, **64**, 814–817.

Taylor, O.C., Stephens, E.R., Darley, E.F., & Cardiff, E.A. (1960). Effect of air-borne oxidants on leaves of pinto bean and petunia. *Proc. Am. Soc. Hort. Sci.* **75**, 435–444.

Taylor, O.C., Duggar, W.M., Cardiff, E.A., & Darley, E.F. (1961). Interactions of light and atmospheric photochemical products (smog) within plants. *Nature*, **192**, 814–816.

Temple, P.J. (1989). Effects of oxidant air pollution in vegetation of Joshua Tree National Monument. *Environ. Pollut.* (in press).

Temple, P., & Bisessar, S. (1979). Response of white bean to bacterial blight, ozone, and antioxidant protection in the field. *Phytopathology*, **69**, 101–103.

Temple, P.J., Taylor, O.C., & Benoit, C.F. (1985). Cotton yield responses of ozone as mediated by soil moisture and evapotranspiration. *J. Environ. Qual.*, **14**, 55–60.

Temple, P.J., Taylor, O.C., & Benoit, C.F. (1986). Yield response of head lettuce (*Lactuca sativa* L.) to ozone. *Environ. Exp. Bot.*, **26**, 53–58.

Temple, P.J., Taylor, O.C., Benoit, C.F., Lennox, R.W., & Reagan, C.A. (1988). Yield response of alfalfa as influenced by O_3 and soil moisture stress. *J. Environ. Qual.*, **17**, 108–113.

Thomas, M.D., & Alther, E.W. (1966). The effects of fluorine on plants. In *Handbook of Experimental Pharmacology*, 20, pp. 231–306. Pharmacology of Fluorides, Springer-Verlag, New York.

Thomas, M.D., & Hendricks, R.H. (1956). Effect of air pollution on plants. Section 9. In R.L. Magill, F.R. Holden, & C. Ackley (eds), *Air Pollution Handbook*, McGraw Hill, New York.

Thomas, M.D., Hendricks, R.H., & Hill, G.R. (1950). Sulfur metabolism of plants. Effects of sulfur dioxide on vegetation. *Ind. Engl. Chem.* **42**, 2231–2235.

Thomas, M.D., Henricks, R.H., & Hill, G.R. (1952). Effects of air pollution on plants. In L.C. McCabe (ed.), *Air Pollution*, pp. 41–65, McGraw-Hill, New York.

Thompson, C.R. (1968). Effects of air pollutants on lemons and navel oranges. *Calif. Agric.*, **22**, 2–3.

Thompson, C.R., & Taylor, O.C. (1969). Effects of air pollution on growth, leaf drop, fruit drop and yield of citrus trees. *Environ. Sci. Technol.* 3, 934–940.

Thompson, C.R., Taylor, O.C., Thomas, M.D., & Ivie, J.O. (1967). Effects of air pollutants on apparent photosynthesis and water use by citrus trees. *Environ. Sci. Tech.*, **1**, 614–650.

Thompson, C.R., Katz, G., & Hensel, E. (1972). Effects of ambient levels of ozone on navel oranges. *Environ. Sci. Technol.* **6**, 1014–1016.

Thompson, C.R., Katz, G., & Lennox, R.W. (1980). Effects of SO_2 and/or NO_2 on native plants of the Mojave Desert and Eastern Mojave-Colorado desert. *J. Air Poll. Contr. Assoc.*, **30**, 1304–1309.

Thompson, C.R., Olszyk, D.M., Kats, A., Bytnerowicz, A., Dawson, P.J., & Wolf, J.W. (1984). Effects of ozone on sulfur dioxide on annual plants of the Mojave desert. *J. Air Poll. Contr. Assoc.*, **34**, 1017–1022.

Thompson, M.A. (1981). Tree rings and air pollution: A case study of *Pinus monophylla* growing in east-central Nevada *Environ. Pollut. (Ser. A.)*, **26**, 251–266.

Thomson, W.W. (1975). Effects of air pollutants on plant ultrastructure. In J.B. Mudd & T.T. Kozlowski (eds), *Responses of Plants to Air Pollution*, pp. 179–194, Academic Press, New York.

Thomson, W.W., Duggar, W.M., & Palmer, R.L. (1965). Effects of peroxyacetyl nitrate on ultrastructure of chloroplasts. *Bot. Gaz. (Chicago)*, **126**, 66–72.

Thomson, W.W., Duggar, W.M., Jr, & Palmer, R.L. (1966). Effect of ozone on the fine structure of the palisade parenchyma of bean leaves. *Can. J. Bot.*, **44**, 1677–1682.

Thomson, W.W., Nagahashi, J., & Platt, K. (1974). Further observations on the effects of ozone on the ultrastructure of leaf tissue. In M. Duggar, (ed.), *Air Pollution Effects on Plant Growth*, pp. 83–93, ACS Ser. 3, Am. Chem. Soc., Washington, DC.

Ting, I.P., & Duggar, W.M., Jr. (1968). Factors affecting ozone sensitivity and susceptibility of cotton plants. *J. Air Poll. Contr. Assoc.* **18**, 810–813.

Ting, I.P., & Duggar W.M., Jr. (1971). Factors affecting ozone sensitivity and susceptibility of cotton plants. *J. Air Poll. Contr. Assoc.*, **18**, 310–313.

Tingey, D.T., & Reinert, R.A. (1975). The effect of ozone and sulfur dioxide singly and in combination on plant growth. *Environ. Pollut.*, **9**, 117–125.

Tingey, D.T., & Taylor, G.E. Jr (1982). Variation in plant response to ozone: A conceptual model of physiological events. In M.D. Unsworth & D.P. Ormrod (eds), *Effects of Gaseous Air Pollutants in Agriculture and Horticulture*, pp. 111–138, Butterworths, London.

Tingey, D.T., Heck, W.W., & Reinert, R.A. (1971). Effect of low concentrations of ozone and sulfur dioxide on foliage, growth and yield of radish. *J. Am. Soc. Hort. Sci.*, **96**, 369–371.

Tingey, D.T., Dunning, J.A., & Jividen, G.M. (1973). Radish root growth reduced by acute ozone exposure. In *Proc. 3rd Intern. Clean Air Congress*, pp. A154–156, VD1-Verlag, GmbH, Dusseldorf.

Tingey, D.T., Reinert, R.A., Wickliff, C., & Heck, W.W. (1973). Chronic ozone or sulfur dioxide exposures, or both effect the early vegetation growth of soybean. *Can. J. Plant Sci.*, **53**, 875–879.

Tingey, D.T., Fites, R.C., & Wicliff, C. (1976a). Differential foliar sensitivity of soya bean cultivars to ozone associated with differential enzyme activities. *Physiol. Plant.*, **37**, 69–72.

Tingey, D.T., Wilhour, R.G., & Standley, C. (1976b). The effect of chronic ozone exposures on the metabolic content of ponderosa pine seedlings. *For. Sci.*, **22**, 234–241.

Tingey, D.T., Thutt, G.L., Gumpertz, M.L., & Hagsett, W.E. (1982). Plant water stress influences ozone sensitivity of bean plants. *Agric. Environ.*, **7**, 243–254.

Todd, G.W., Middleton, J.T., & Brewer, R.F. (1956). Effects of air pollutants. *Calif. Agric.*, **9**, 7–8, 14.

Toyama, S. (1976). Effects of photochemical oxidants on plant organelles. *Sholubutsu Boeki*, **30**, 223–229.

Treshow, M. (1965a). Evaluation of vegetation injury as an air quality criterion. *J. Air Poll. Contr. Assoc.*, **15**, 266–269.

Treshow, M. (1965b). Response of some pathogenic fungi to sodium fluoride. *Mycologia*, **57**, 216–221.

Treshow, M. (1968). The impact of air pollutants on plant populations. *Phytopathology*, **58**, 1108–1113.

Treshow, M. (1970a). *Environment and Plant Response*, McGraw Hill, New York. 422 pp.

Treshow, M. (1970b). Ozone damage to plants. *Environ. Pollut.*, **1**, 155–161.

Treshow, M. (1971). Fluorides as air pollutants affecting plants. *Ann. Rev. Phytopathol.*, **9**, 21–44.

Treshow, M. (1975). Interaction of air pollutant and plant disease. In J.B. Mudd & T.T. Kozlowski (eds), *Responses of Plants to Air Pollution*, pp. 307–334, Academic Press, London.

Treshow, M. (1976). *Environment and Plant Response*. McGraw Hill, New York.

Treshow, M. (1980a). Interactions of air pollutants and plant disease. In P.R. Miller (ed.), *Effects of Air Pollutants on Mediterranean and Temperate Forest Ecosystems*, pp. 103–109, Proc. Symp., Riverside, CA, June 22–27, 256 pp.

Treshow, M. (1980b). Pollution effects on plant distribution. *Environ. Conserv.*, **7**, 279–286.

Treshow, M. (1983). Effects of acid deposition on forest vegetation: Physiological and biochemical. *Conf. on Acid Rain and Forest Resources*, Quebec, June 14–17.

Treshow, M. (Ed.) (1984). *Air Pollution and Plant Life*. Wiley, Chichester. xii + 486 pp., illus.

Treshow, M. (1988). The role of pollutants in the productivity of the forest ecosystem of developed and developing countries: An overview. *Internat. Symp. on Plants and Pollutants*, 22–28 August, Izmir, Turkey,

Treshow, M., & Allan, J. (1979). Annual variation in the dynamics of a woodland plant community. *Environ. Conserv.*, **6**, 231–236.

Treshow, M., & Allan, J. (1985). Uncertainties associated with the assessment of vegetation. *Environ. Management*, **9**, 471–478.

Trehow, M., & Deumling, D. (1985). Symptomatological look at forest decline in Germany. *Perspect. Environ. Bot.*, **1**, 261–269.

Treshow, M., & Pack, M.R. (1970). Fluoride. In J.S. Jacobson & A.C. Hill (eds), *Recognition of Air Pollution Injury to Vegetation: A Pictorial Atlas*, Air Pollution Control Association, Pittsburgh.

Treshow, M., & Stewart, D. (1973). Ozone sensitivity of plants in natural communities. *Biol. Conserv.*, **5**, 209–214.

Treshow, M., Anderson, F.K., & Harner, F.M. (1967). Responses of Douglas fir to elevated atmospheric fluoride. *Forest Sci.*, **13**, 114–120.

Treshow, M., Harner, F.M., Price, H.E., & Kormelink, J.R. (1969). Effects of ozone on growth, lipid metabolism and sporulation of fungi. *Phytopathology*, **59**, 1223–1225.

Treshow, M., Sutherland, E.K., & Bennett, J. P. (1986). Tolerance and susceptibility to air pollution, a new direction in sampling strategy. In *Proc. Int. Symp. Ecol. Aspects of Tree-Ring Analysis*, pp. 392–400, Conf., Aug. 17–21.

Tveite, B. (1984). Til na ingen forverra tilstand i Norst skog [Until now no decreased vitality in Norwegian forests]. *Norsk. Skogbr.*, **30**, 16.

US Department of Agriculture (1982). *Agricultural Statistics, 1981*, US Govt. Printing Office, Washington, DC.

US Department of Commerce (1986). *Monthly Climatic Data for the World*, NOAA. Env. Data Serv. Ashville, NC.

US Department of Health, Education and Welfare (1966). *The Effects of Air Pollution*, US Public Health Service Publ. 1556, 18 pp.

US Environmental Protection Agency (1983). *Can We Delay a Greenhouse Warming?* Washington, DC.

US Environmental Protection Agency (1984). *National Air Quality and Emissions Trends Report, 1983*, Research Triangle Park, NC, US EPA Office of Air Quality Planning and Standards, EPA Rept. No. EPA 450/4-B4-029.

US Environmental Protection Agency (1986). *Air Quality Criteria for Ozone and Other Photochemical Oxidants*, Environ. Criteria and Assessments Office, Research Triangle Park, NC, EPA-600/8-84-020.

US Environmental Protection Agency (1988). *National Air Quality and Emissions Trends Report, 1986*, Office of Air Quality Planning and Standards, Research Triangle Park, NC, EPA 450/4-88-001, 214 pp.

Ullmann, C. (1896). Landwirtschaftliche Beliage der Norddentaschen. *Allgemeinen Zeitung*, No. 17.

Ulrich, B., Mayer, R., & Khanna, T.K. (1979). *Deposition von Luftverunreinigungen und ihre Auswirkungen in Waldökosystemen in Solling*. Shrifter aus der Forstl. Fak. der Univ. Göttingen u. d. Niederachs. Forstl. Versuchsanst. **58**, 291 pp.

Ulrich, B., Mayer, R., & Khanna, T.K. (1980). Chemical changes due to acid precipitation in a losses-derived soil in central Europe. *Soil Sci.*, **130**, 193–199.

Unsworth, M.H., Biscoe, P.V., & Black, V. (1976). Analysis of gas exchange between plants and polluted atmospheres. In T.A. Mansfield (ed.), *Effects of Air Pollutants on Plants*, pp. 4–16, Cambridge University Press, Cambridge.

US Environmental Protection Agency (1981). *Preliminary Assessment of Health and Welfare Effects Associated with Nitrogen Oxides for Standard Setting Purposes.* Rev. draft staff paper, USEPA, Research Triangle Park, NC. 68 pp. + appendix.

Van der Eerden, L.J.M. (1982). Toxicity of ammonia to plants. *Agric. Environ.*, **7**, 223–235.

Von Eckstein, D., Aniol, R.W., & Bauch, J. (1983). Dendroclimatological investigations on fir dieback. *Eur. J. For. Pathol.*, **13**, 279.

Wagner, F. (1981) Ausmass und Verlauf des Tannensterbens in Ost Bayern von 1975–1980. *Forstwiss. Centralbl.*, **100**, 148–160.

Wang, T.H., Lin, C.S., Wu, C., Liao, C., & Line, H. (1949). Fluorine content of Eukien teas. *Food Res.*, **14**, 98–103.

Wanta, R.C., Moreland, W.B., & Heggestad, H.E. (1961). Tropospheric ozone: An air pollution problem arising in the Washington D.C. metropolitan area. *Monthly Weather Rev.*, **89**, 289–296.

Watson, R. (1988). *Executive Summary, Ozone Trends panel Report*, NASA Headquarters, Washington, DC.

Weinstein, L.H. (1961). Effects of atmospheric fluoride on metabolic constituents of tomato and bean leaves. *Contrib. Boyce Thompson Inst.*, **21**, 215–231.

Weinstein, L.H. (1977). Fluoride and plant life. *J. Occup. Med.*, **19**, 49–78.

Weinstein, L.H., & Alscher-Herman, R. (1982). Physiological responses of plants to fluorine. In M.H. Unsworth & D.P. Ormrod (eds), *Effects of Gaseous Air Pollution in Agriculture and Horticulture*, pp. 139–176, Butterworths, London.

Wellburn, A.R. (1982). Effects of SO_2 and NO_2 on metabolic function. In M.H. Unsworth & D.P. Ormrod (eds), *Effects of Gaseous Air Pollution in Agriculture and Horticulture*, pp. 169–188, Butterworths, London.

Wellburn, A. (1988). *Air Pollution and Acid Rain*, Longren, New York, 274 pp.

Wellburn, A.R., Majernik, O., & Wellburn, F.A.M. (1972). Effects of SO_2 and NO_2 polluted air upon the ultrastructure of chloroplasts. *Environ. Pollut.*, **3**, 37–49.

Went, F.W. (1955). Air pollution. *Scientific American*, **192**, 62–70, 72.

Whitmore, M.E., Freer-Smith, P.H., & Davis, T. (1982). Some effects of low concentrations of SO_2 and/or NO_2 on the growth of grasses and poplar. In M.H. Unsworth & D.P. Ormrod (eds), *Effects of Gaseous Air Pollution in Agriculture and Horticulture*, pp. 483–485, Butterworths, London.

WHO (1987). *Air Quality Guidelines for Euope.* WHO Regional Public Series No. 23, 426 pp.

Wilhour, R.G. (1970). Influence of ozone on white ash (*Fraxinus americana* L.) *Pennsylvania State University Center Air Environ. Studies Publication*, **188**, 71.

Williams, W.T., Brady, M., & Wilson, S.C. (1977). Air pollution damage to the forests of Sierra Nevada mountains of California. *J. Air Poll. Contr. Assoc.*, **27**, 230–234.

Winkler, P. (1982). Zur Trendentwicklung des pH. Wertes des Niederschlages in Mitteleuropa. *Z. Pflanzen. Boden.*, **145**, 576–585.

Wolff, G.T., & Lioy, P.J. (1980). Development of an ozone river associated with synoptic-scale episodes in the eastern United States. *Environ. Sci. Technol.*, **14**, 1257–1260.

Wolff, G.T., Kelley, N.A., & Ferman, M.A. (1982). Source regions of summertime ozone and haze episodes in the eastern United States. *Water Air Soil Pollut.*, **18**, 65–81.

Wood, C.W., & Nash, T.N., III (1976). Copper smelter effluent effects on the Sonoran desert vegetation. *Ecology*, **57**, 1311–1316.

Woodman, J.N., & Cowling, E.B. (1987). Airborne chemicals and forest health. *Environ. Sci. Tech.*, **21**, 120–126.

Woodwell, G.M. (1970). Effects of pollution on the structure and physiology of ecosystems. *Science*, **168**, 1101–1107.

Woodwell, G.M., Hobbie, J.E., Houghton, R.A., Melillo, J.M., Moore, B., Peterson, B.J., & Shaver, G.R. (1983). Global deforestation: Contribution to atmospheric carbon dioxide. *Science*, **222**, 1081–1086.

World Health Organization (1961). *Air Pollution*, Columbia University Press, NY, 442 pp.

Wukasch, R., & Hofstra, G. (1977). Ozone and *Botrytis* spp. Interaction in onion leaf dieback field studies. *J. Ann. Soc. Hort. Sci.*, **102**, 543–546.

Yamazoe, F., & Mayumi, H. (1977). Vegetation injury from interaction of mixed air pollutants. *Proc. Internat. Clean Air Congr.* **4**, Tokyo, Japan.

Yarwood, C.E., & Middleton, J.T. (1954). Smog injury and rust infection. *Plant Physiol.*, **29**, 393–395.

Zahn, R. (1970). The effect on plants of a combination of subacute and toxic sulfur dioxide doses. *Staub*, **30**, 20–32.

Ziegler, I. (1973). Effect of sulphite on phosphoenolpyruvate carboxylase and malate formation in extracts of *Zea mays*. *Phytochemistry*, **12**, 1027–1030.

Ziegler, I. (1975a). The effect of SO_2 pollution on plant metabolism. *Residue Rev.*, **56**, 79–105.

Ziegler, I. (1975b). The effect of air polluting gases on plant metabolism. *Environ. Qual. Saf.*, **2**, 182–208.

Zimmerman, P.W., & Hitchcock, A.E. (1956). Susceptibility of plants to hydrofluoric acid and sulfur dioxide gases. *Contrib. Boyce Thompson. Inst.*, **18**, 263–279.

Zoettl, H.W., & Huettl, R.F. (1986). Nutrient supply and forest decline in southwest Germany. *Water Air Soil Pollut.*, **31**, 449–462.

Zwiazek, J.J. (1987). Fluoride-induced and draught-induced structural alterations of mesophyll and guard cells in cotyledons of Jackpine (*Pinus banksiana*). *Can. J. Bot.*, **65**, 2310–2317.

APPENDIX A

Scientific Names
of Plants

Common Name	Scientific Name
Ailanthus	*Ailanthus altissima* (Mill.) Swingle
Alfalfa	*Medicago sativa* L.
Alder	*Alnus tenuifolia* Nutt.
Annual bluegrass	*Poa annua* L.
Apple	*Malus sylvestris* Mill.
Apricot	*Prunus armeniaca* L.
Arborvitae	*Thuja* sp. L.
oriental	*T. orientalis* L.
Ash	*Fraxinus* sp. L.
European	*Fraxinus excelsior* L.
European mountain	*Sorbus aucuparia* L.
Green	*F. pennsylvanica* Marsh var. *lanceolata Sang*
red	*F. pennsylvanica* Marsh.
Aspen	*Populus* sp. L.
quaking	*P. tremuloides* Michx.
Aster	*Aster* sp. L.
Avocado	*Persea americana* Mill.
Azalea	*Rhododendron* sp. L.
Bachelor's button	*Centaurea cyanus* L.
Barberry	*Berberis* sp. L.
Barley	*Hordeum vulgare* L.
Bean	*Phaseolus* sp. L.
bush	*P. vulgaris* L. var. *humilis* Alef.
lima	*P. limensis* Macf.
navy	*P. vulgaris* L.
pinto	*P. vulgaris* L.
pole	*P. vulgaris* L.
red kidney	*P. vulgaris* L.
snap	*P. vulgaris* L.
Beech	*Fagus* L
European	*Fagus sylvatica* L.
Beet	*Beta* L.
sugar	*Beta vulgaris* L.

Bindweed	*Convolvulus arvensis*
Birch	*Betula* sp. L.
European white	*Betula pendula* Roth
paper	*B. papyrifera* Marsh.
silver	*B. verrucosa* Chrh.
yellow	*B. lutea* Michx.
Black medic	*Medicago lupulia* L.
Bluegrass	*Poa pratensis* L.
Box elder	*Acer negundo* L.
Boysenberry	*Rubus* L.
Bridalwreath	*Spiraea prunifolia* Sieb. & Zucc.
Brittle brush	*Encelia farinosa* Gray
Broccoli	*Brassica oleracea* L. var. *italica* Plench
Brussels sprout	*B. oleracea* L. var. *gemmifera* Zenher Plench.
Buckeye	
California	*Aesculus californica* Nutt.
Ohio	*A. glabra* Wild.
Buckwheat	*Fagopyrum esculentum* Maench.
Burrow weed	*Ambrosia dumosa* Payne
Cabbage	*Brassica oleracea* L. var. *capitata*. L.
Chinese	*B. pekinensis* Rupr.
Calendula	*Calendula* sp. L.
Camellia	*Camellia* sp. L.
Careless weed	*Amaranthus palmeri* L.
Carnation	*Dianthus* sp. L.
Carrot	*Daucus carota* L.
Catalpa	*Catalpa bignoniodies* Walt.
Cauliflower	*Brassica botrytis* L.
Cedar, red	*Juniperus virginiana* L.
Celery	*Apium graveolens* L.
Chard, Swiss	*Beta vulgaris* L. var. *cicla* L.
Cheeseweed	*Malva neglecta* Wallr.
Cherry, Bing	*Prunus avium* L. var. Bing
Cherry, Bitter	*P. emarginata* L.
Chickweed	*Cerastium* L.
Chokecherry	*Prunus virginiana*
Chrysanthemum	*Chrysanthemum* L.
Citrus	*Citrus* sp. L.
Clover	
ladino, white	*Trifolium repens* L.
red	*T. pratense* L.
white sweet	*Melilotus alba* Medic.
yellow sweet	*Melilotus officinalis* (L.) Pallas
Coleus	*Coleus* sp. Lour.
Corn	*Zea mays* L.
Cosmos	*Cosmos* Cav.
Cotoneaster, rock	*Cotoneaster horizontalis* Decne.
Cotoneaster, spreading	*C. divanicata* Reh & Wils.
Cotton	*Gossypium* sp. L.
Cottonwood	*Populus* sp. L.
Creosote brush	*Covillea glutinosa* Rydb.

Cucumber	*Cucumis sativus* L.
Curly dock	*Rumex crispus* L.
Currant, black	*Ribes petiolare* Dougl.
Cyclamen	*Cyclamen* sp. L.
Dandelion	*Taraxacum officinale* Web.
Desert willow	*Chilopsis linearis* Sweet
Dock	*Rumex* sp. L.
curly-leaf	*R. crispus* L.
Dogwood	*Cornus stolonifera* Michx.
Elderberry	*Sambucus* L.
Elm, American	*Ulmus americana* L.
Chinese	*U. parvifolia* Jocq.
Siberian	*U. pumila* L.
Endive	*Cichorium endiva* L.
Euonymus	*Euonymus* sp. L.
False Solomon's Seal	*Smilicina* sp.
Fescue	*Festuca* L.
Fir	*Abies* sp. Mill.
balsam	*Abies balsamea* Mill.
Douglas	*Pseudotsuga menziesii* Franco
Greek	*Abies cephalonica* Loud.
Grand	*A. grandis* Lindl
Sacred	*A. religiosa* Lindl.
Silver	*A. alba* Mill.
Spanish	*A. pinsapo* Boiss.
Subalpine	*A. lasiocarpa* Nutt.
white	*A. concolor* Hoopes
Firethorn	*Pyracantha coccinea* Depp.
Fleabane	*Erigeron canadensis* Cronq.
Forsythia	*Forsythia* sp. Vahl.
Four O'Clock	*Mirabilis jalapa* L.
Fuchsia	*Fuchsia* L.
Garden cress	*Lepidium sativum* L.
Gifblaar	*Dishapetalum cymosum*
Gladiolus	*Gladiolus* sp. L.
Globe mallow	*Sphaeralcea grossularifolca* Rydb.,
Goosefoot, nettle-leaf	*Chenopodium* L.
Grape	*Vitis* sp. L.
Grapefruit	*Citrus paradisi* Macf.
Grass	
bent	*Agoseris* sp.
blue	*Poa pratensis* L.
brome	*Bromus* sp.
cheat	*Bromus tectorum* L.
crab	*Digitaria ischanemum* Schreb.
galleta	*Hilaria jamesii* Berth.
Indian rice	*Oryzopsis hymenoides* Ricker.

Johnson	*Sorghum halepense* Pers.
orchard	*Dactylis glomerata* L.
pine	*Calamogrostis rubescens* Buckl.
rye	*Lolium* sp. L.
salt	*Distichlis stricta* Rydb.
wheat	*Agropyron* sp. Beauv.
Ground cherry	*Physalis* sp. L.
Gum, black	*Nyssa sylvatica* Pepperidge
Hawthorn	*Crataegus* L.
Heather	*Erica tetralix* L.
Hemlock, Eastern	*Tsuga canadensis* Carr.
Holly	*Ilex* sp. L.
American	*Ilex opaca* Ait.
Mahonia	*Mahonia aquifolium* Nutt.
Hypericum	*Hypericum* sp. L.
Incense cedar	*Libocedrus decurrens* Torr.
Iris	*Iris* sp. L.
Ivy, English	*Hedera helix* L.
Japanese box	*Buxus microphylla* Sieb. & Zucc.
Jerusalum-Cherry	*Solanum pseudo-capsicum* L.
Juniper, alligater	*Juniperus pachyphlaea* Torr.
Chinese	*J. chinensis* L.
mountain	*J. communis* L.
Rocky Mountain	*J. scopulorum* Sarg.
Utah	*J. osteosperma* Little
Kentia palm	*Howe belmoreana* Becc.
Lamb's quarter	*Chenopodium album* L.
Larch, European	*Larix decidua* Mill.
Japanese	*L. leptolepis* Gord.
Laurel, mt.	*Kalmia latifolia* L.
Lemon	*Citrus lemon* Burm.
Lettuce, head	*Lactuca capitata* L.
prickly	*L. serriola* L.
Lilac	*Syringa vulgaris* L.
Linden, American	*Tilia americana* L.
Linden, little leaf	*T. cordata* Mill.
Locust, black	*Robinia pseudoacacia* L.
honey	*Gleditsia triacanthos* L.
Maize	*Zea mays* L.
Maple, hedge	*Acer campestre* L.
Norway	*A. platanoides* L.
mountain	*A. glabrum* Torr
red	*A. rubrum* L.
silver or white	*A. saccharinum* L.
sugar or rock	*A. saccharum* Marsh.

Milkweed	*Asclepias syriaca* L.
Mimulus	*Mimulus* sp. L.
Mint	*Mentha* sp. L.
Mock orange	*Philadelphus coronarius* L.
Morning-glory	*Ipomoea purpurea* Lam.
Mountain laurel	*Ceanothus sanguineus* Pursh.
Mulberry	*Morus rubra* L.
Muskmelon	*Cucumis melo* L.
Mustard	*Brassica* sp. L.
Narcissus	*Narcissus* L.
Nightshade	*Solanum* L.
Oak, black	*Quercus kelloggii* Newl.
Oak, Bur	*Quercus macrocarpa* Michx.
Chestnut	*Q. montana* Willd.
Gambel, Scrub	*Q. gambelii* Nutt.
pin	*Q. palustris* Muenchh.
red	*Q. rubra* L.
scarlet	*Q. coccinea* Muench.
white	*Q. alba* L.
Okra	*Hibiscus esculentus*
Onion	*Allium* sp. L.
Orange	*Citrus sinensis* Osbeck.
Orange, mock	*Philadelphus* sp.
Oregon grape	*Mahonia repens* Don.
Parsley	*Petroselinum crispum* Nym.
Peach	*Prunus persica* Batsch.
Peanut	*Archis hypogaea* L.
Pear, Bartlett	*Pyrus communis* L. var. Bartlett
pea	*Pisum sativum* L.
Pepper	*Piper nigrum* L.
Petunia	*Petunia* sp. Juss.
Wild Petunia	*Ruellia strepens* Pursh.
Pigweed	*Amaranthus retroflexus* L.
Piggyback plant	*Tiarella menziesii* Pursh.
Pine	*Pinus* sp. L
Austrian	*Pinus nigra* Arnold.
black	*P. nigra* Arnold.
Chinese	*P. tabulaeformis* Curr.
Coulter	*P. coulteri* D. Don
digger	*P. sabiniana* Dougl.
fox	*P. aristata* Engelm.
Eastern white	*P. strobus* L.
jack	*P. banksiana* Lamb.
Jeffrey	*P. jeffreyi* A. Murr.
knobcone	*P. attenuata* Lemm.
limber	*P. flexilis* James
lodgepole	*P. contorta* Loud.
pinyon	*P. edulis* Engelm.
(single leaf)	*P. monophylla* T. & F.

pitch	*P. rigida* Mill.
ponderosa	*P. ponderosa* Laws
red	*P. resinosa* Ait.
shortleaf	*P. echinata* Mill.
Scots, Scotch	*P. sylvestris* L.
sugar	*P. brutins* Tan.
Virginia	*P. virginiana* Mill.
Western white	*P. monticola* Dougl.
Plane tree	*Platanus* sp. L.
Plaintain	*Musa paradisiaca* L.
Poison Ivy	*Rhus radicans* L.
Poplar, Caroline	*Populus deltoides* Michx.
white	*P. alba* L.
yellow	*Liriodendron tulipifera* L.
Potato	*Solanum tuborosum* L.
Primrose	*Primula* sp. L.
Privet, Lodense	*Ligustrum vulgare* var. *pyramidale*
Prune	*Prunus domestica* L.
Pumpkin	*Cucurbita pepo* L.
Radish	*Raphanus sativus* L.
Ragweed	*Ambrosia* sp. L.
Ranunculus	*Ranunculus* sp. L.
Raspberry	*Rubus idaeus* L.
Redbud	*Cercis canadensis* L.
Redwood	*Sequoia sempervirens* Endl.
Dawn	*Metasequoia glyptostroboides*, Hu & Cheng.
Rhododendron	*Rhododendron* L.
Rhubarb	*Rheum rhaponticum* L.
Rice	*Oryza sativa* L.
Romaine	*Lactuca longifolia* Lam.
Rye	*Secale cereale.* L.
Saltbush	*Atriplex canescens* Nutt.
Safflower	*Carthamus tinctorius* L.
Sequoia, giant	*Sequoiadendron giganteum*, Buchholz.
Serviceberry	*Amelanchier alnifolia* Nutt.
Smartweed	*Polygonum acre* H.B.K.
Snapdragon	*Antirrhinum majus* L.
Sorghum	*Sorghum vulgare* Pers.
Soya bean	*Glycine max* Merr.
Spice bush	*Linera benzoin* B.
Spinach	*Spinacia oleracea* L.
Spruce	*Picea* sp. A. Dietr.
black	*P. mariana* BSP.
Black hills	*P. glauca* var. *densata* Bailey
Colorado blue	*P. pungens* Engelm.
Englemann	*P. engelmanii* Parry
Norway	*P. abies* Karst.
red	*P. rubens* Sarg.
western white	*P. glauca* Voss.
white	*P. abies* Karst

Squash, summer	*Cucurbita pepo* L. var. *melopepo* Alef.
Strawberry	*Fragaria* sp. L.
Sunflower	*Helianthus* sp. L.
Sweetgum	*Liquidambar styraciflua* L.
Sweet potato	*Ipomoea batatas* Lam.
Sweet pea	*Lathyrus odoratus* L.
Swiss chard	*Beta vulgaris* L.
Sycamore, American	*Platanus occidentalis* L.
Tea plants	*Thea sinensis* L.
Thimbleberry	*Rubus parviflorus* L.
Timothy	*Phleum pratense* L.
Tobacco	*Nicotiana tabacum* L.
Tomato	*Lycopersicon esculentum* Mill.
Tulip	*Tulipa gesneriana* L.
Turnip	*Brassica rapa* L.
Velvet weed	*Gaura parviflora* Dougl.
Verbena	*Verbena canadensis* Britt.
Viburnum, Burkwood	*Viburnum burkwoodii* Busk.
Violet	*Viola* sp. L.
African	*Saintpaulia ionantha* Wendl.
Wheat	*Triticum aestivum* L.
Willow	*Salix* sp. L.
Yellow buckeye	*Aesculus octandra* Marsh.
Yew, Japanese	*Taxus cuspidata* Sieb. & Zucc.
Zinnia	*Zinnia* sp.

Conversion of Concentration Units for Gaseous Pollutants Between ppm by Volume and μg/m³ Mass Loadings

Atmospheric concentrations for gaseous pollutants are often reported in parts per million (ppm) or parts per billion (ppb, in which 'billion' means 10^9 or 'thousand-million') by volume. The unit 'ppm' means 'microliters per liter (μl/l), which is 10^{-6} liters of pollutant per liter of air. Atmospheric particulate concentrations are usually reported by mass or weight per unit of air volume, usually as micrograms per cubic meter of air ($\mu g/m^3 = 10^{-6}g/m^3$) but also sometimes as milligrams per cubic meter ($mg/m^3 = 10^{-3}g/m^3$). Sometimes particulate concentrations are reported as numbers of particles in given size ranges per cubic centimeter or other volume of air. This is done from actual counts using laser particle counters or by scanning electron microscopy of suitable filter samples.

Many researchers and regulatory agencies use the mass loading method for gaseous pollutants too. The conversion from ppm or ppb pollutant volumes to $\mu g/m^3$ is easily done by multiplying the pollutant volume by the molecular weight and a conversion factor involving the molar volume, temperature, and pressure. The reverse conversion is also easy. One should remember, however, that converting observations of particulate data to the gaseous ppm form is usually meaningless and inadmissible. Most airborne particles are chemically complex substances, many of which have no gaseous counterpart at ambient temperatures and pressures.

The standard temperature and pressure (STP) for reporting air pollutant concentrations are 25 °C (= 298.15 °K) and one atmosphere of pressure. In air pollution literature, all atmospheric concentrations of pollutants are reported with reference to these standard conditions, for both gaseous and particulate pollutants, and in both volume and weight methods. This means that one can convert reported gaseous pollutant concentrations freely between ppm or ppb and $\mu g/m^3$ without needing to know the temperature and pressure of the original observations.

To convert from $\mu g/m^3$ to ppm, use formula (1). If the data are in mg/m^3, rewrite them in $\mu g/m^3$ form before applying the formula by moving the decimal three places to the right. If the data are for any other temperature or pressure than 25 °C and 1 atm (STP), the STP correction factor in the square brackets must be applied. At STP, this correction factor is equal to 1 and may be disregarded.

To convert from ppm to $\mu g/m^3$, use formula (2). If the data are in ppb, rewrite them in ppm form before applying the formula by moving the decimal three places to the left.

(1)
$$\text{ppm} = \left(\frac{\mu g}{m^3}\right)\left(\frac{0.024478}{M}\right)\left[\left(\frac{P_o}{P}\right)\left(\frac{T}{T_o}\right)\right]$$

(2)
$$\frac{\mu g}{m^3} = (\text{ppm})40.853M\left[\left(\frac{P}{P_o}\right)\left(\frac{T_o}{T}\right)\right]$$

in which M = molecular weight of the pollutant (g/mole),

P = ambient pressure at the time of the observation (atm),

P_o = pressure at standard conditions (1 atmosphere),

T = ambient temperature in degrees Kelvin (°C + 273.15)

T_o = temperature at standard conditions (25 °C + 273.15 = 298.15 °K)

The constant in formula (1) is from the volume of one mole of gas at 25 °C and 1 atm (24.4781 liters, called the 'molar volume') combined with a units conversion factor of 10^{-3} m^3/l. It is obtained from the combined gas laws equation, $PV = nRT$, by solving for V (volume) at 25 °C and 1 atm, using the gas constant $R = 0.0821$ (l)(atm)/(g/mole)(°K). The constant in formula (2) is the reciprocal of the one in formula (1).

Molecular weights of selected gaseous air pollutants

Gas	Formula	Molecular weight
Ammonia	NH_3	17.032
Carbon monoxide	CO	28.011
Chlorine	Cl_2	70.914
Hydrogen chloride	HCl	36.465
Hydrogen fluoride	HF	20.008
Hydrogen sulfide	H_2S	34.082
Methane	CH_4	16.043
Nitric oxide	NO	30.008
Nitrogen dioxide	NO_2	46.008
Nitrous oxide	N_2O	44.016
Ozone	O_3	48.000
Peroxyacetyl nitrate (PAN)	$CH_3COO_2NO_2$	121.054
Sulfur dioxide	SO_2	64.066
Sulfur trioxide	SO_3	80.066

Index